$5.95

GEM TRAILS
OF
NEW MEXICO

James R. Mitchell

Book Co.

3677 SAN GABRIEL RIVER PARKWAY

PICO RIVERA, CALIF. 90660

Library of Congress Catalogue Card Number 87-81646

ISBN 0-935182-24-1

Contributing Editor: *MARLENE LEE*
Cover Design: *BRETT PALMER*
Maps: *JEAN HAMMOND*

Note: Due to the possibility of misinterpretation of information, *Gem Trails of New Mexico*, its author publisher and all other persons directly or indirectly involved with this publication assume no responsibility for accidents, injury or any losses by individuals or groups using this publication.

In rough terrain and hazardous areas all persons are advised to be aware of possible changes due to man or nature that can occur along the gem trails.

INTRODUCTION

New Mexico is the sixth ranked mineral producing state in the entire nation, boasting vast supplies of zinc, lead, gold, silver, magnesium, potash, uranium, helium, copper, coal, gypsum and salt, as well as large quantities of oil and natural gas. Obviously, with such a high degree of mineralization, there are many fine places for rockhounds to pursue their hobby, and the best are described on the following pages.

Be advised that some of the sites are located on the dumps of old abandoned mines. Do not, under any circumstances, enter the shafts and always be cautious when exploring the surrounding regions. There are often hidden tunnels, rotten ground, and pits, as well as rusty nails, broken glass and discarded chemicals; all of which create potential hazards.

Each of the locations listed in the revised *Gem Trails of New Mexico* was visited and carefully checked to verify mineral availability and current collecting status. A few of the spots are on private property and a fee is charged. That information, as well as the fee at time of publication, is noted in the text. DO NOT ASSUME THAT THIS GUIDE GIVES PERMISSION TO COLLECT! Land status changes frequently. If you have a suspicion that a particular site is no longer open, be sure to ascertain the status before trespassing. If nothing can be determined locally, land ownership information is available at the County Recorder's office.

Most the areas are easy to get to, but road conditions do change. Severe weather can make good roads very rough, and very rough roads impassible, even with four-wheel drive. You must decide for yourself which of them your particular vehicle is capable of traveling.

These sites are situated in landscapes as full of variety as the minerals themselves. Generally, however, New Mexico is cold during the winter and hot during the summer, making the best time to visit either the spring or fall. If you must go during the winter or summer months, be sure to take proper clothing and other supplies. During those months, weather conditions can drastically change in a very short period of time. In the winter, a mountain covered with warm sunlight in the morning might be blanketed with snow by afternoon. During the summer, a parched desert can be under water in minutes during a severe thunder storm.

When venturing into some of the more remote areas, it is a good idea to take extra drinking water, foul weather clothing, and possibly even some food just in case you get delayed or stuck. If you take time to properly plan your collecting trip and make sure your vehicle is in good working order, the gem fields listed on the following pages will provide you and your family with outstanding minerals and many memorable experiences.

James R. Mitchell

Key Map to Section Locations

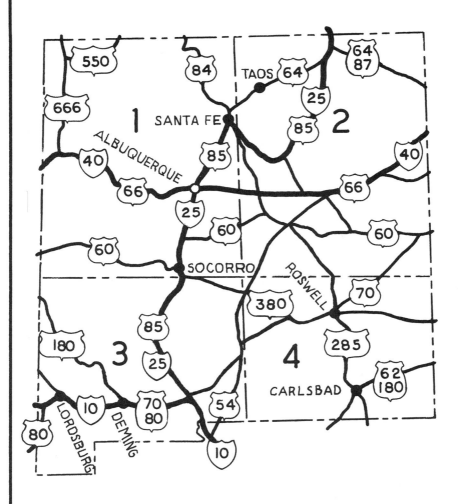

TABLE OF CONTENTS

GRANTS

Countless Apache tears, as well as small topaz and garnet crystals, can be found a short distance north of Grants. To get to the collecting site, take Lobo Canyon Road (Highway 547) north from the Lobo Canyon Shopping Center, five and three-tenths miles, as shown on the map. The site is easy to spot, being where the massive cliffs get closest to the highway. At the given mileage, there are some ruts on the left, which provide a good place to park.

The tears are scattered all over the terrain throughout the foothills for quite a distance. The topaz and garnet, however, are much tougher to find, generally being concealed in gas cavities of the rhyolite blocks which can be seen crumbling down from the higher cliffs.

The crystals are small, but some are sizeable enough to produce fine faceted stones. It takes some time and work to obtain these gemstones, so be willing to allow sufficient time to make the effort worthwhile. The most productive method is to use a sledge hammer and chisel to break up any suspect rhyolite and then carefully examine all gas cavities for included crystals. When they are found, carefully pop them out with a nail or other pointed object and put them somewhere where they won't get lost.

Parking area near cliffs

Grants

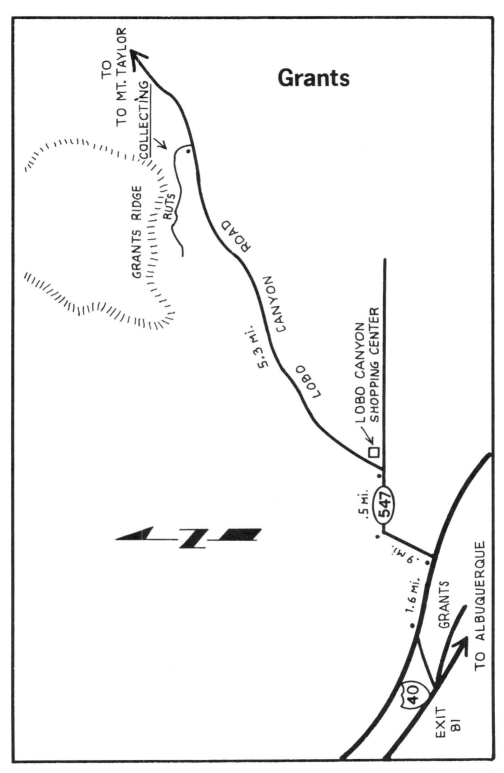

RIO PUERCO

The region on both sides of the Rio Puerco River, near where Interstate 40 crosses, is littered with colorful agate, jasper, petrified wood and Apache tears. To get there, drive about nineteen miles west from Albuquerque to the Rio Puerco exit (Exit #140) and go north to the Frontage Road. At that point, turn right, cross the river, and you will be at the southern edge of this extensive and productive location.

Take any of the ruts leading north from the pavement to the gravel topped hills only a short distance away. The collecting is done on those hills which extend for quite a distance. Most of what can be found is somewhat small, being primarily suitable only for tumbling. Large chunks, however, are not too hard to obtain.

If you aren't satisfied with what you get at the first stop, simply move on a short distance and try again. The agate and jasper occur in a wide range of colors, while the wood is primarily grey and brown, most being suitable only for display as is rather than cut and polished.

There are more roads on the west side of the river making those gravel topped hills more accessible, but that is part of the Laguna Indian Reservation. If you want to collect there, get permission before going in.

Collecting on a gravel topped hill

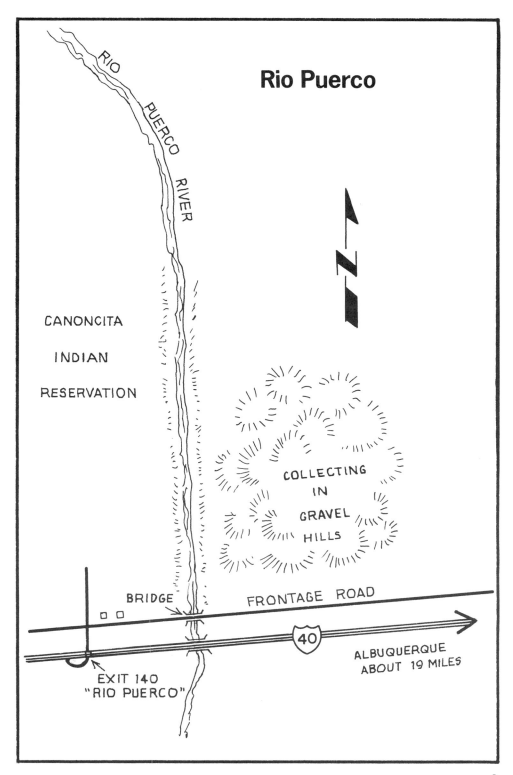

Rio Puerco

RIO PUERCO RIVER

CANONCITA INDIAN RESERVATION

COLLECTING IN GRAVEL HILLS

BRIDGE

FRONTAGE ROAD

40

ALBUQUERQUE ABOUT 19 MILES

EXIT 140 "RIO PUERCO"

APACHE CREEK

Large amounts of colorful agate can be found in Lee Russell Canyon only a short distance north of Apache Creek. To get to this highly productive rockhounding area, start at the Apache Creek store, which is on Highway 12, about 11 miles northeast of Reserve. From there, go north on State Highway 32 four and nine-tenths miles to where a road can be seen leading through a locked gate on the left. Park well off the pavement, hike around the gate, and head across the little river. The narrow belt of land on both sides of the waterway is private property, but collectors are allowed to walk on the Forest Service road to the National Forest boundary, which is about one-fourth of a mile from the pavement.

The mouth of Lee Russell Canyon is hidden from view at first, but after crossing the creek it can easily be seen directly ahead. Once inside the National Forest, collecting is allowed anywhere, and you should be able to spot lots of colorful agate after having gone only a very short distance.

The best material, however, is obtained about three to five miles further along, but that involves a long hike. If you choose to go the entire way, be sure you are physically capable and take along some water and a snack.

At first the canyon is wide and rocky, and lots of agate can be found along the way, giving a good idea of what is ahead. As you proceed, different colors and types of inclusions will be encountered, and the canyon becomes deeper and more rugged. Finally, the road climbs out on the right onto Turkey Flat and Elk Horn Park where agate displaying an incredible variety of colors and patterns can be found; this is the prime area. Be sure you don't fill your collecting bag too full, however, since it is a long hike back to where you parked.

Hiking the trail into Lee Russell Canyon

Apache Creek
Agate

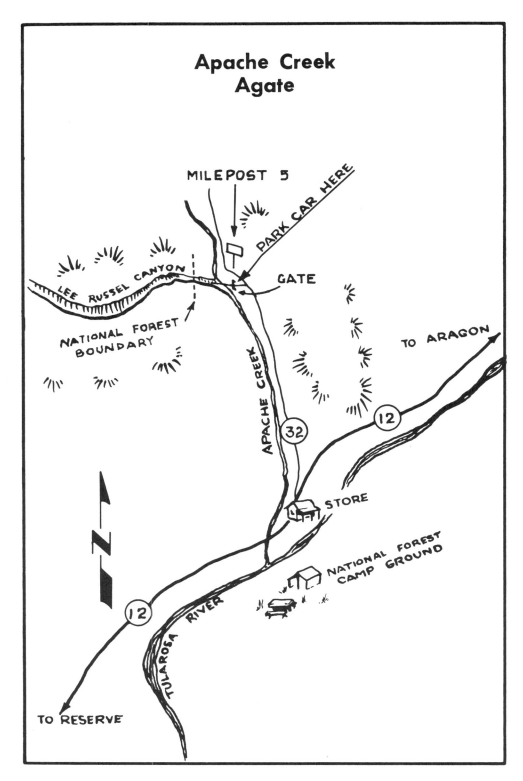

LUNA AREA

Nice banded agate, closely resembling that found in Brazil, can be picked up in the forest near Luna. Most of the washes in the region offer good collecting possibilities, but there are a few spots in particular which are especially productive. The first is reached by taking Highway 180 west from Luna about four and three-tenths miles to a bridge which crosses the San Francisco River. The agate can be found by hiking either north or south along the river from the bridge.

Amethyst crystals, in geodes, can be obtained on the ridge north of Highway 180 about two miles from town. The geodes are not easy to find, but it is a fun site to explore. Complete geodes generally are found only by digging. Look for fragments in the loose soil for clues as to where to start.

Another good agate area is in the creek north of town and west of the rodeo grounds where large chunks have washed down from the surrounding hills.

Southeast of Luna, near Highway 180, you can pick up some clear agate containing brilliant black dendritic inclusions. The location is on the south side of the highway between the logging road shown on the map and the San Francisco Mountains.

Amethyst crystals, flower-like clusters of clear quartz crystals, and blue banded agate are all found ten miles southeast of Luna on the north side of Highway 180.

It should be noted that Luna is an ideal summer collecting area, situated at an elevation of nearly 6000 feet, providing pleasant cool days for hunting and cold nights for sleeping. This is generally not a suitable location for winter collecting, however.

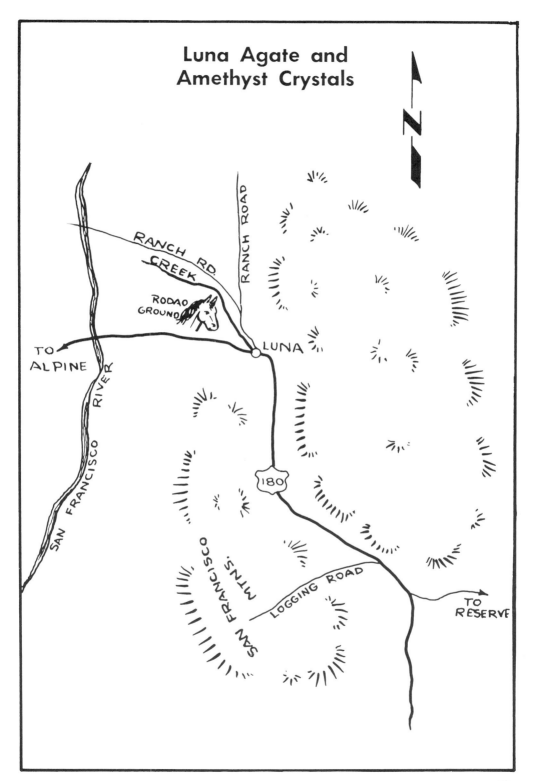

Luna Agate and Amethyst Crystals

RED HILL AREA

Red Hill is only a service station situated on the side of Highway 60. It, however, is the prime landmark for finding the road to this collecting area, so be sure to find it on a map before starting out. From Red Hill, take the ranch road which heads north from the highway only a short distance east of the gas station. This road is not difficult to spot because of a cattle guard just off the highway. Once on that fairly well graded dirt road, proceed north about six miles to a metal water tank, which will be on the right. From there, continue another two and six-tenths miles to another cattle guard. Just before reaching it, a dim set of ruts will be seen on the left going around a power pole. You should follow them as they lead into a valley. After continuing about one more mile, you will be in the center of the collecting area. Nice agate and jasper can be found by inspecting the volcanic rocks for quite a distance in all directions. The material is not as plentiful here as at many other places, but some chunks of the agate are filled with moss-like inclusions, making them real prizes and well worth the extra effort needed to find them.

This is a remote area, and if you choose to visit, be sure to go in with someone else or let friends know where you are going.

A specimen of moss agate

Red Hill Area

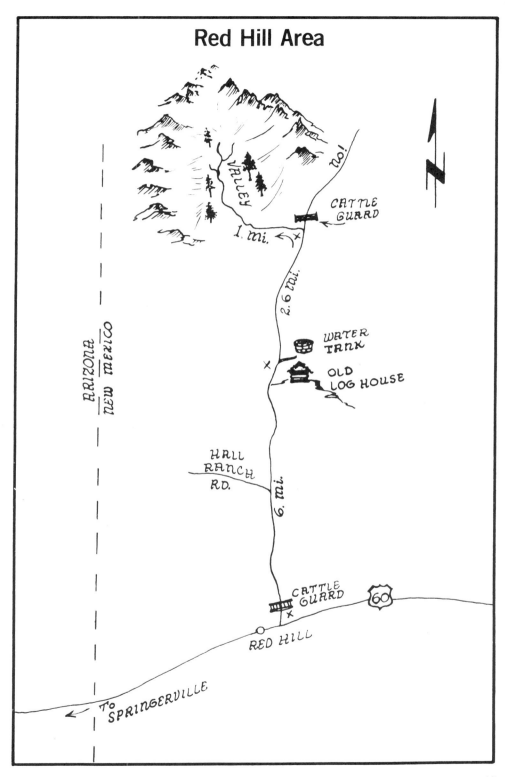

VALLEY

no!

CATTLE GUARD

1 mi.

2.6 mi.

ARIZONA / NEW MEXICO

WATER TANK

OLD LOG HOUSE

HALL RANCH RD.

6 mi.

CATTLE GUARD

60

RED HILL

To SPRINGERVILLE

MAGDALENA

One of New Mexico's most prized minerals is the vivid blue smithsonite from the old Kelly Mine, a short distance south of Magdalena. The Kelly is currently closed to amateur collectors but there are other mines in the region which are open to those willing to pay a small fee.

The best known of the Magdalena fee sites are the dumps at the Nit and North Graphic. Both are operated by William Dobson, and he can be contacted at Bill's Gem & Mineral Shop adjacent to the Ponderosa Restaurant in Magdalena. At the shop, you will be given detailed instructions as to how to get to the dumps and, in addition, you will receive a key needed to open a locked gate on the road leading to both mines. While at the shop, you can also view a selection of typical specimens from the two localities. The fee at time of publication is $2.00 per person per day with a twenty pound limit.

At the two mines, you will be able to find nice specimens of azurite, barite, pyrite, bornite, iron ores and even some of the prized smithsonite. Mr. Dobson is very helpful and will be glad to answer any questions you might have.

For additionial information, be sure to stop at Tony's Rock Shop located on the left side of the road about three-fourths of a mile from Highway 60 on the way to Kelly. The owner worked at the Kelly Mine when it was operating, and, more recently, has served as caretaker. He can give you a wealth of additional information about local mining, and while in the shop, you will be able to see even more specimens.

The old Kelly Mine

Magdalena
Mineral Specimens

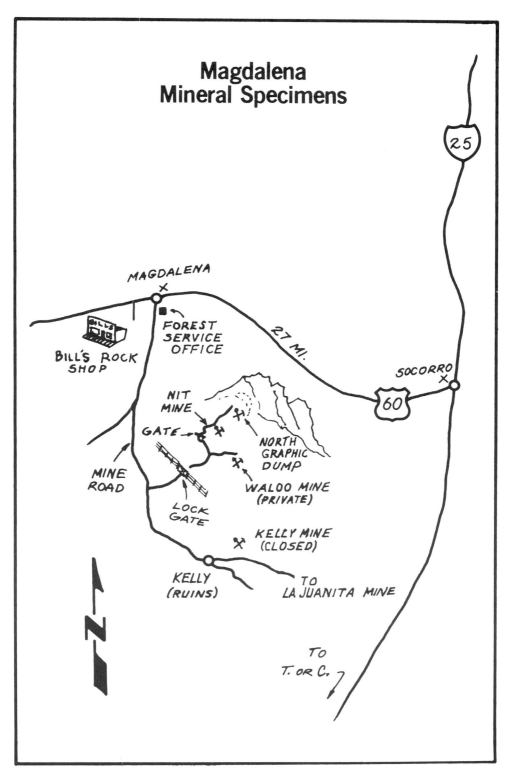

RILEY

This is a very remote site, so be certain your vehicle is in good working order before heading in. The center of the collecting area is the old ghost town of Riley, situated on the northern bank of Rio Salado about twenty miles north of Magdalena.

Take the paved road leading north from Highway 60 in the middle of town. Just before the fairgrounds, bear right onto the graded dirt road (Forest Service Road #354) and go nineteen and one-half miles. At that point, the main road forks. If you go left, the tracks pass a farm and head across the river. Be certain your vehicle can make it, especially if there is much water flowing. Otherwise, take the right fork to river's edge and walk.

When hiking, be careful climbing down the steep bank to the river bed. In addition, as is the case when driving across, don't go if lots of water is flowing. Usually there is little, if any, and the walk is easy.

The old town of Riley can be seen on the opposite banks, and when there do not disturb any of the buildings or the cemetery. Former residents still visit from time to time, and the old town is very important to them.

Fossils are found in the shale on the hill behind what appears to be the old school house. White barite crystals can be dug in pits encountered in and around the town as well as in much of the loose soil at the base of the shale. Most of the barite pits are now filled in due to erosion of the soft surrounding soil, but the pick and shovel work needed to clean them out is usually rewarded with some interesting intertwined specimens.

Lots of agate and jasper can be picked up on the hills surrounding the town, as well as in the town itself. In addition, more can be found on the opposite side of the river near where you parked if you walked across. Most are fairly small and great for tumbling. Some, however, are sizeable enough for making cabochons. All tend to be good quality, displaying a wide variety of colors and internal patterns.

The ghost town of Riley across the Rio Salado

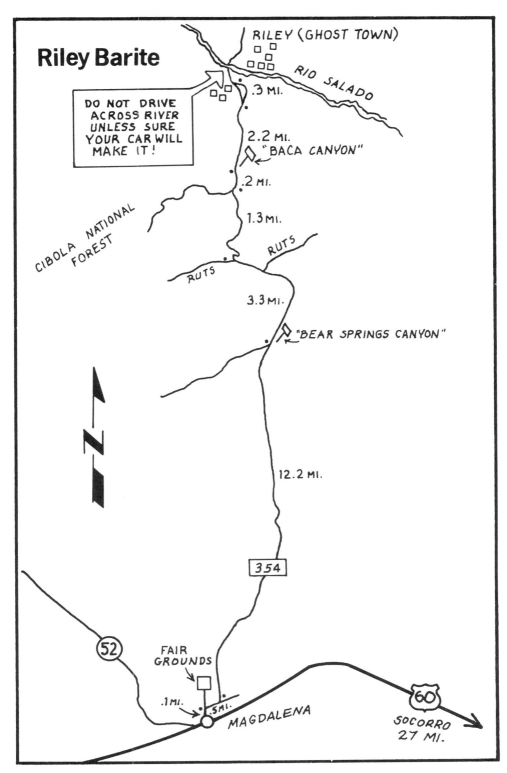

Riley Barite

RILEY (GHOST TOWN)

RIO SALADO

DO NOT DRIVE ACROSS RIVER UNLESS SURE YOUR CAR WILL MAKE IT!

.3 MI.

2.2 MI.
"BACA CANYON"

.2 MI.

1.3 MI.

RUTS

RUTS

CIBOLA NATIONAL FOREST

3.3 MI.

"BEAR SPRINGS CANYON"

N

12.2 MI.

354

52

FAIR GROUNDS

.1 MI.

.5 MI.

MAGDALENA

60

SOCORRO 27 MI.

SOCORRO FOSSILS

At one time in the distant geological past, the area near Socorro was very different than it is now. To the east was the Carlile Sea, stretching for miles across the lowlands in the region of the now prominent Los Pinos Mountains.

That body of water dried up long ago, but evidence of its existence in the form of fossilized shells and other sea life can be found throughout this part of the state. The site discussed here offers collectors a good opportunity to gather a selection of those fossils.

To get there, follow the instructions on the map, parking well off the pavement at the head of the wash. From there, simply walk about 100 yards up the wash and proceed cross-country to the flat-topped hills on the right. The fossils are in the blocky rock on top. As you go, be on the lookout for interesting chert nodules and rhombohedral calcite crystals which are scattered about on the lowlands.

At the summit, virtually all of the rock contains fossils, the most prized being the often sizeable and well preserved crinoid stems. In addition, there are worms, seaweed, and a huge variety of shells. Use a rock pick and a chisel to remove portions of the rock in hopes of exposing good specimens.

The variety of what can be found in a given chunk is incredible. You could spend hours simply trying to identify all the ancient life forms you will encounter. Be sure to try a number of different spots along the ridge, and don't hesitate to examine the boulders that have broken away and fallen part way down the hill. In fact, you will often be able to gather a better selection if you concentrate your efforts on the loose rock rather than the tough primary deposit. Your time and effort will be more productive, since the smaller pieces are generally easier to split.

Crinoid stems at fossil site

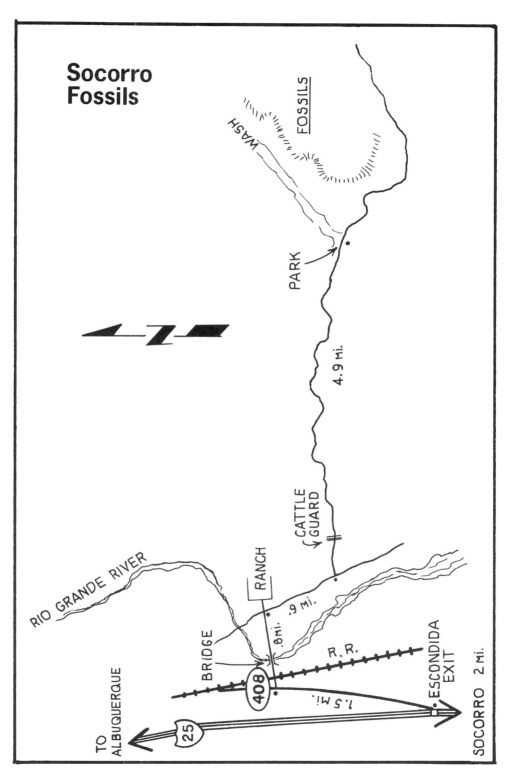

Socorro
Fossils

FOSSILS

WASH

PARK

4.9 mi.

CATTLE GUARD

RIO GRANDE RIVER

RANCH

.9 mi.

.8 mi.

BRIDGE

R.R.

408

1.5 mi.

ESCONDIDA EXIT

SOCORRO 2 mi.

TO ALBUQUERQUE

25

CHUPADERA MOUNTAINS

The Chupadera Mountains, a short distance south of Socorro, are well known for the manganese that has been mined there through the years. One such prospect, which has long been abandoned, offers collectors an opportunity to find a variety of minerals, including chalcedony, agate, botryoidal hematite and velvet psilomelane.

To get to this interesting and productive collecting site, simply head south from Socorro on Interstate 25 about ten miles to the San Antonio turnoff (Highway 380). Instead of going into San Antonio, go across the interstate and bear south across the cattle guard. At that point, the pavement ends and there is a Forest Service sign giving mileage to various points including the Cowart Ranch. Only one-tenth of a mile further you should turn right and continue five and three-tenths miles toward the mountains. When you approach the given mileage, it will be easy to see the colossal quarry on the hillside.

The best collecting seems to be on the northern side of the ridge, about eight-tenths of a mile further, and it is suggested you go there first.

The agate and jasper are scattered throughout the quarries and surrounding hills; some is very colorful and worth picking up. Much of the hematite is bubbly and, if large enough, highly desirable for display in mineral collections. The velvet botryoidal psilomelane is why most collectors visit, though. It fills the cracks and cavities of the quarry rock, and the finest specimens are obtained by splitting suspect stones along obvious fractures in hopes they will open into psilomelane filled voids. If you choose to work on the quarry walls, be very careful, since some of the rock is unstable and potentially dangerous. By the time you finish here, your hands and clothing will be covered with black manganese ores; that is the price you must pay.

Botryoidal psilomelane from manganese quarry

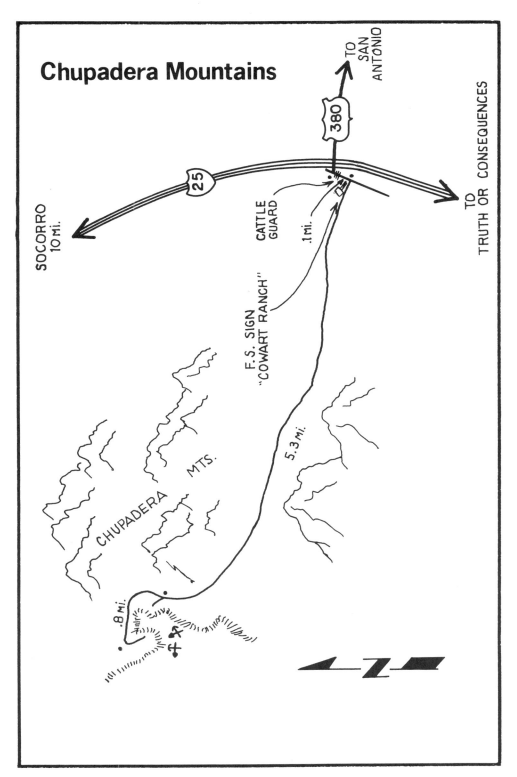

Chupadera Mountains

SOCORRO
10 Mi.

25

380

TO
SAN
ANTONIO

TO
TRUTH OR CONSEQUENCES

CATTLE
GUARD

.1 mi.

F.S. SIGN
"COWART RANCH"

5.3 mi.

CHUPADERA MTS.

.8 mi.

CHAVEZ CANYON

This is a fairly remote area and requires some hiking and climbing to get to most of the collecting spots. For those reasons it is a site primarily of interest to rockhounds who enjoy exploring and working for what they get.

Take the Highway 380 exit from Interstate 25, about ten miles south of Socorro. Instead of following Highway 380, go west across the interstate and then south on the pole line road. Take the second turn to the right, which is about four-tenths of a mile from where you left the pavement. Proceed three more miles, and then go right another one and three-tenths miles on the somewhat rough tracks that lead over the mountain into Chavez Canyon.

Site A is the ridge just across the canyon from where the road enters the wash, and on that ridge one can find rhyolite nodules filled with crystals and/or banded agate. In addition, on the upper regions, there are seams of white calcite and alabaster; some of which fluoresces yellow-orange under an ultraviolet light.

Site B is reached by going up Chavez Wash three-tenths of a mile and then hiking about one-quarter of a mile up the narrow canyon on the right. From there, go right into the intersecting canyon approximately two hundred yards to a brown opalite deposit. Yellow and orange jasper filled with intricate dendrite patterns can also be found here.

Continuing in Chavez Wash an additional four hundred yards will place you next to yet another narrow canyon leading off to the right. A short distance up that canyon is Site C, which boasts agate, jasper and a scattering of chalcedony and opalite.

Most of the hills and washes in this region offer lots of material of interest to rockhounds, so try to explore as much of the area as possible.

Parked near Chavez Wash

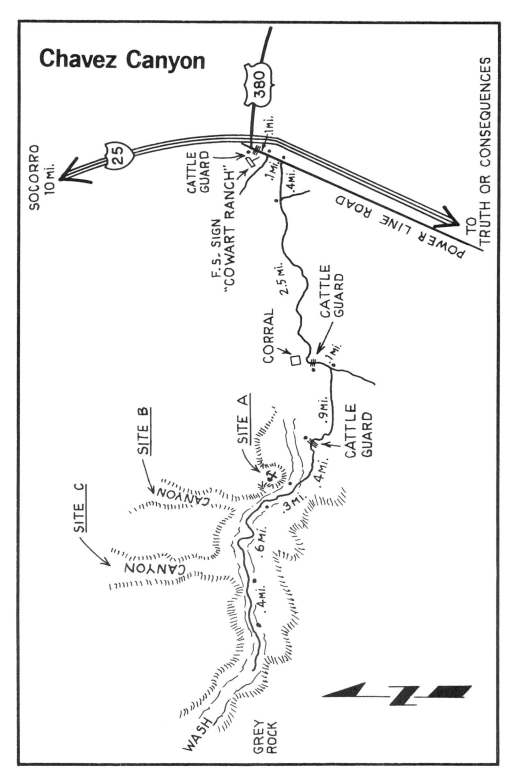

Chavez Canyon

COYOTE AREA

This is one of many outstanding agate locations situated near Pedernal Peak. In fact, so much can be found along the roads in this area that your biggest problem will not be finding sufficient quantities, but, instead, deciding what to keep and what to leave behind. It is tough to sort through such a plentiful supply of colorful agate trying to decide which is the best.

To reach this highly productive spot, take Forest Service Road #316 south from Highway 96 in Coyote, or go west about two miles from town to Forest Service Road #172. Be advised that it is not a good idea to drive on the roads surrounding Pedernal Peak during or immediately after a heavy rain, since they can get very muddy and slippery in places.

Agate can be found scattered on both sides of either road, a short distance after leaving the highway, all the way to the Rio Puerco Campground, as shown on the map. Probably the best way to collect here is to drive a few miles, park and inspect the terrain on both sides of the road. Then, continue a few more miles and repeat the procedure. By doing that, you will have a better sample of everything that the site has to offer. It seems that the agate is more colorful at some spots along the road than at others. Also, the quantities vary considerably from place to place.

The entire loop, as shown on the map, is about twenty-five miles, and some portions of the road are rough. Most rugged vehicles should have no problem, however, if driven carefully.

This is a productive collecting site so be sure to allow sufficient time to adequately explore it.

Apache tears and red jasper from the Rio Puerco area

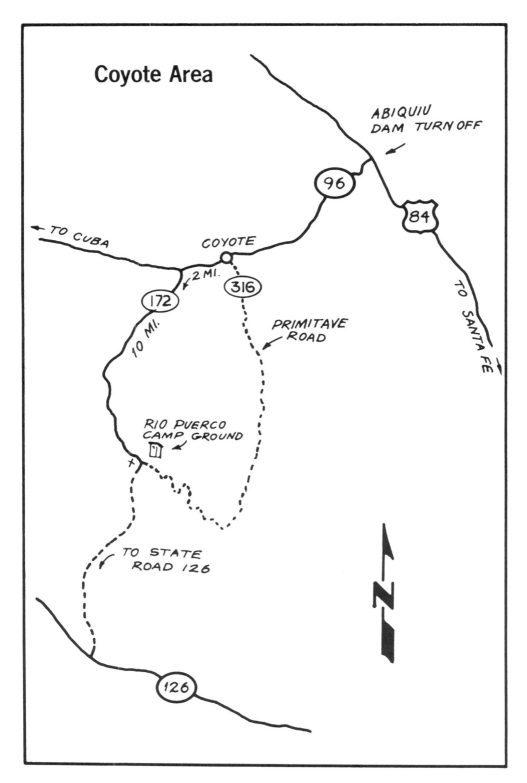

Coyote Area

ABIQUIU DAM TURN OFF

96

84

TO CUBA

COYOTE

2 MI.

172

316

10 MI.

PRIMITAVE ROAD

TO SANTA FE

RIO PUERCO CAMP GROUND

TO STATE ROAD 126

N

126

CUBA AREA

There are two very good sites near Cuba which offer rockhounds a wide variety of collectibles. The first location boasts fine examples of petrified wood, and the second is noted for its nice mineral specimens. To get to the wood location, take Highway 197 southwest from Cuba about four and two-tenths miles to where a dirt road intersects on the right. Go through the gate and follow that road toward the mesa. Specimen grade wood can be found in all directions as you approach the hills, but most is somewhat porous and not suitable for polishing. It seems that better grade wood is more plentiful as you get nearer to the mesa. Most of the top quality cutting material, though, is somewhat small, being suitable only for tumbling or smaller cabochons.

The second Cuba location is an old copper mine and is reached by returning to town and heading east on Highway 126 as it winds it way into the mountains. After having gone nine and eight-tenths miles, you will see some ruts leading off to the left which you should follow. From there, it gets fairly steep and rough in places, so be sure your vehicle is capable of traveling on such a road. If you have any doubts, simply park off Highway 126 and hike the rest of the way.

The mine dump is located about one-half mile from the highway and offers very colorful specimens of blue-green chrysocolla and malachite. In fact, nice pieces can also be found along the road to the mine, so be on the lookout as you approach.

Cuba Area

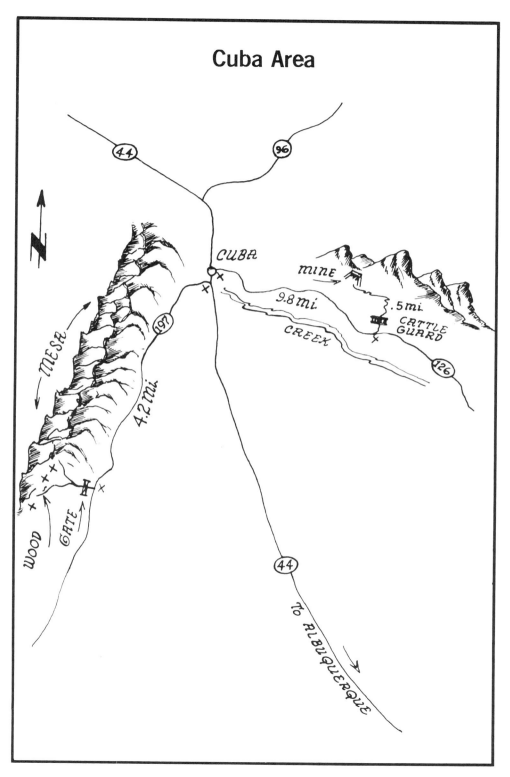

LOS LUNAS

Tons of top quality agate, jasper and petrified wood can be found scattered throughout an extensive area not far from Los Lunas.

To get to this most prolific collecting area, take Interstate 25 to the Los Lunas exit and instead of going to town head west on Highway 6 four and six-tenths miles. The turn to the collecting site on the south is just before the highway crosses the railroad tracks. If you reach the tracks, it will be necessary to double back and try again.

Follow the fairly good dirt road three and two-tenths miles, bypass the railroad maintenance buildings and take the right fork, as illustrated on the map. Continue another five and nine-tenths miles, being careful to stay on the main road since there will be some crossroads along the way.

At the given mileage, there is a small water tank on the left side of the road. To the right, extending for miles, are some shallow hills and eroded mounds tapering down into the huge arroyo. It is in those mounds where you will find a nearly unlimited supply of extraordinarily colorful agate, jasper and petrified wood, as well as lots of Apache tears.

Most of what can be found here is somewhat small, but good fist size chunks are not uncommon. Much of the jasper is in shades of red and orange, while the agate tends to be white, with seemingly unlimited inclusion patterns and colors. The real prize, though, are the small chocolate brown pieces of petrified wood inundated with brilliant areas of red.

Collecting area for agate, jasper and petrified wood

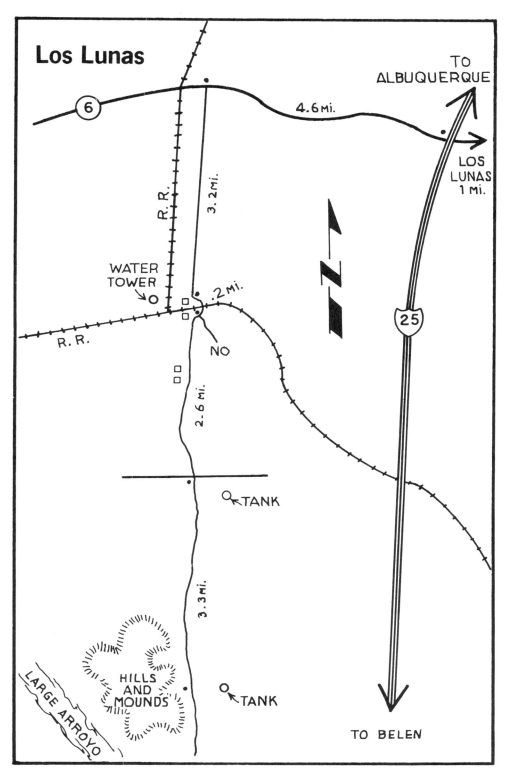

Los Lunas

WEST OF ALBUQUERQUE

It is hard to believe that a productive agate and jasper field is located on the outskirts of a town the size of Albuquerque. That is the case, though, and if you would like to see for yourself, simply take Interstate 40 west from town about five miles to the 98th Street Exit (Exit #153). Head north from the highway toward the obvious black ridge that marks the edge of the collecting site.

The pavement ends after going only one-half mile and as you near the ridge, it gets very sandy. If your vehicle cannot drive through loose sand, simply park as close as you feel safe going and walk the rest of the way.

The hills and mounds just south of the ridge, extending for quite a distance east and west, are full of colorful jasper, agate and occasionally petrified wood. Walk along the lower slopes of the volcanic flow to find the collecting material, being sure to take enough time to find the best this site has to offer. The jasper is generally found in shades of yellow, red, orange and sometimes in combination with each other. The agate is usually either clear or white, but as was the case with the jasper, every now and then some outstanding, multicolored pieces will be found. The wood is brown, and some is filled with streaks of bright orange and yellow, making them highly desirable for polishing.

Walking the lower slopes of an old volcanic flow

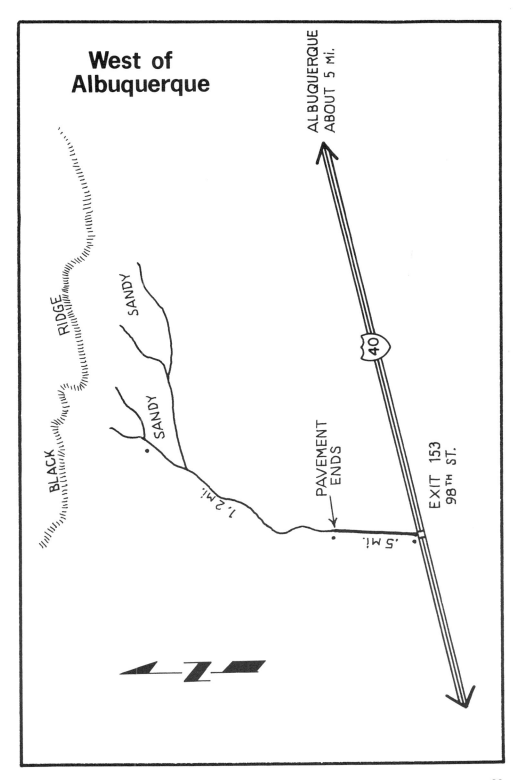

West of Albuquerque

ALBUQUERQUE ABOUT 5 Mi.

BLACK RIDGE

SANDY

SANDY

1.2 mi.

PAVEMENT ENDS

.5 mi.

40

EXIT 153
98TH ST.

33

PETACA AREA

The pegmatites between Petaca and Vallecitos have long been known to prospectors and mineral collectors. The area is primarily noted for its fine specimens of mica, green beryl, amazonite, black columbite, garnet, smoky quartz, schist and beautiful pink feldspar. The New Mexico Bureau of Mines and Mineral Resources, however, lists over fifty different minerals that-can be found in this relatively condensed area.

The hills are filled with dumps, and all offer collectors a chance of finding a variety of minerals. To get to this most interesting and scenic area, start in Petaca, which is about thrity-two miles north of Espanola. From town, get on the dirt road heading over the mountains toward Vallecitos. It is well maintained and should present no problem to most vehicles if free of snow and driven carefully.

After going only a few miles, mine dumps will be seen on both sides of the main road. Don't be afraid to try some of the side roads also, since they frequently lead to other dumps. Just be certain you don't collect on an active claim. There are so many abandoned mines in this area that trespassing on an active claim is unnecessary.

Brilliant white bull quartz can be found scattered about the quarries and throughout the pine trees. It makes very nice garden stones and might also be worth picking up.

Be sure you do not limit your exploration only to the quarries. some large mica and quartz boulders can be found scattered randomly throughout the forest, and many of them offer as good or better mineral specimens than is available at the mines.

Searching one of the quarries

Great Ages of Music

TIME LIFE Music

Demonstration Record

Ⓟ 1984 Time-Life Books Inc

GAR-1 • 33⅓RPM • Stereo

Featuring highlights from:

TCHAIKOVSKY: Romeo and Juliet
Concerto No. 2 • VIVALDI: The Four Seasons
TCHAIKOVSKY: Romeo and Juliet • CHOPIN: Polonaise
J. STRAUSS JR.: Blue Danube (Love Theme) • BEETHOVEN: Piano Concerto No 4
BACH: Brandenburg Concerto No. 6 • BIZET: Carmen Suite
HANDEL: Messiah

TO ORDER, return the order card
or write: Time-Life Music, Time & Life Bldg.,
Chicago, Ill. 60611

Made in U K by Lyntone

LYN 15975

PLACE
COIN HERE IF
SLIMDISC
SLIPS

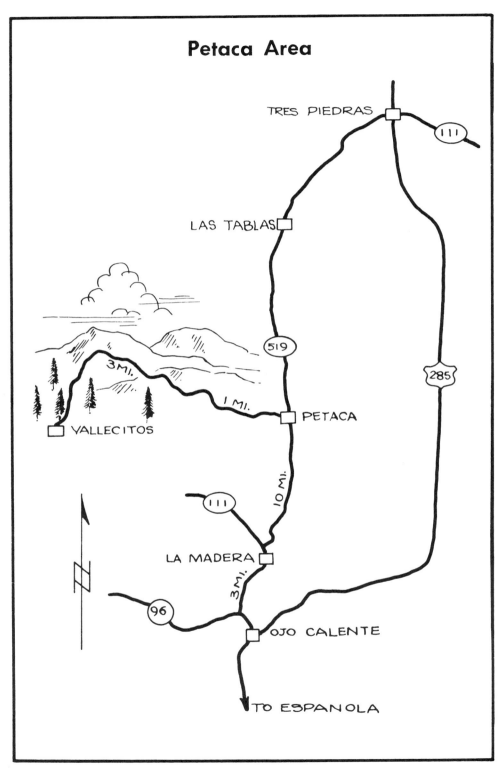

Petaca Area

TRES PIEDRAS

111

LAS TABLAS

519

3 MI.

1 MI.

PETACA

VALLECITOS

10 MI.

111

285

LA MADERA

3 MI.

96

OJO CALENTE

TO ESPANOLA

35

JEMEZ MOUNTAIN AREA

An abundance of minerals, beautiful scenery, and magnificent panoramas have helped make the Jemez Mountain region a favorite rockhounding area for a large number of New Mexico collectors. This fascinating locality begins on Highway 4 on the west and extends over fifty miles to Highway 85 on the east. Something of interest is encountered nearly every mile along the way. There are numerous places to camp, and the road is not bad even though it gets steep and narrow in a few places.

The first location, labeled Site A on the map, features sparkling Apache tears and obsidian. The tears are obtained by digging in the soft, grey perlite hills which can be seen approximately eight-tenths of a mile northwest of the Bear Springs Ranger Cabin. Most of the tears polish nicely and a few will even display faint stars, making them real prizes.

The region just across the main road from the ranger cabin comprises Site B. There you can find agate nodules; most of which are not too spectacular, however. Some have crystal interiors, though, and are nice for display. The best places to look are in the stream beds and other areas of erosion.

About five miles east of the ranger cabin is La Jara Canyon. A road heads north into the canyon and jasper can be found there. The best material seems to be in the first small canyon leading off the main one to the left, Site C, as illustrated on the map. This jasper is worth hunting for, being very similar to the famous Bruno Jasper found in Idaho. The colors tend to be brown, yellow and red, and most is banded or swirled. Just be patient and enjoy the search. If you do so, you will certainly find enough to make the trip worthwhile.

Apache tears and agate from collecting area

Jemez Mountain Area

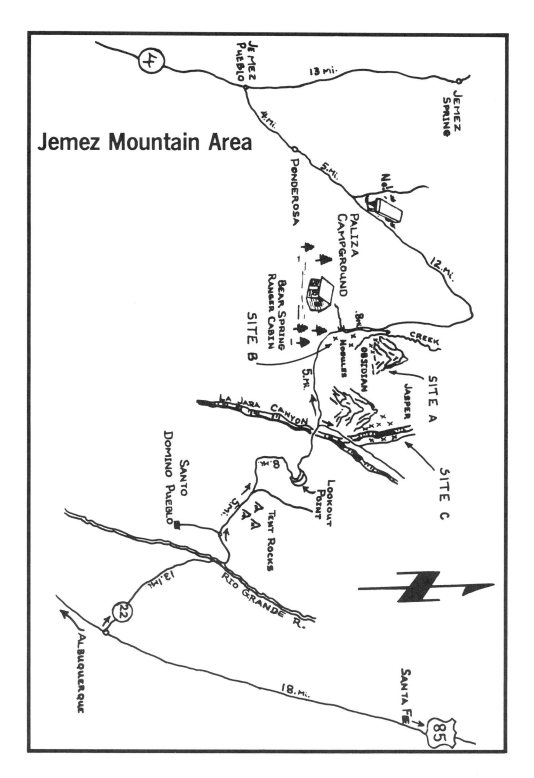

ABIQUIU DAM

This area is about 31 miles northwest of Expanola, just off Highway 96. A dam has been built on the Chama River forming a lake and recreation area, and in the associated picnic area, one can find lots of nice, small agates scattered about. Most of this material was probably brought in when the dam was being constructed, but in any event, some fine material can often be found. A better source, however, is only a short distance away. To get there, take Highway 96 two and six-tenths miles west from where the Abiquiu Dam road intersects. At that point, you will be driving through some low lying hills and should pull off the pavement in a safe spot. That might be difficult if you have a large vehicle or are towing a trailer. In those cases it may be necessary to pull off before reaching the hills. At the given mileage, there is gentle but steep drop from the roadbed, so always be aware of that potential hazard when choosing a suitable place to pull off

To find the agate, simply hike amongst the hills. You will see it everywhere in an amazing variety of patterns and colors, and the quality seems to increase the further you get from the road. The most prized is the shimmering black material filled with spectacular red inclusions. These red and black agates are not overly plentiful but are certainly worth looking for. In addition, there is also lots of white agate; most of which contain a huge variety of inclusions in a multitude of color combinations.

The agate comes in all sizes, ranging from small pebbles to large boulders with most measuring about two to six inches across.

Agate from the Abiquiu Dam area

Abiquiu Dam

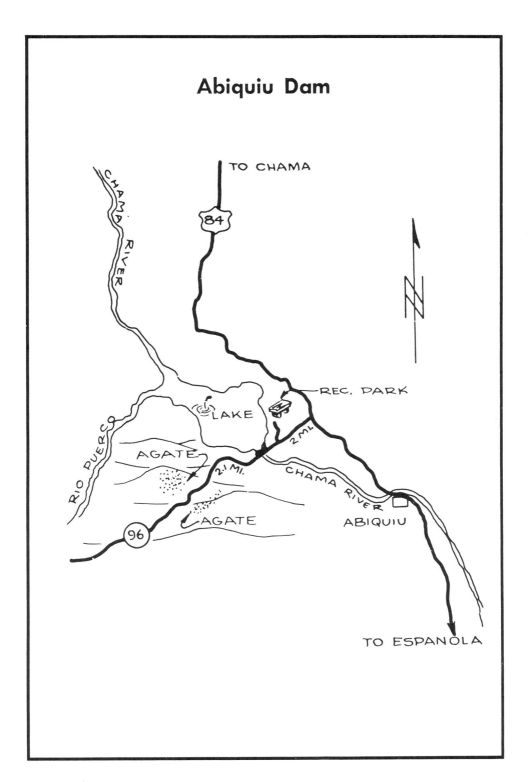

PEDERNAL PEAK

The region surrounding Pedernal Peak is noted for the abundance of top quality milky white agate that can be found there. Especially prized is that which contains beautiful bright red dots and bands. The agate found in this locale is attractive, and it isn't too difficult to procure chunks large enough to produce bookends and spheres.

To get to the most accessible part of this vast collecting area, take Highway 96 west approximately nine miles from where it intersects Highway 84 to Youngsville. From town, take Forest Service Road 100 south toward Pedernal Peak and the Encino Lookout Tower. It should be noted that there are numerous other roads in the vicinity also heading south from the highway toward Pedernal Peak, and most of them will also take you through this most productive agate field. Almost immediately, you will be able to spot agate beside the road, but if you go at least one mile before stopping, you will generally be able to find a better variety of color and inclusion patterns.

After one and two-tenths miles, you will find yourself approaching some unusual weathered sandstone mounds and cliffs. This is a good stop for photographs and more agate. The collecting continues completely through the foothills at least another four miles, so be sure to continue on, stopping a number of times along the way. The concentration varies from spot to spot and so does the quality. The size seems to increase, as is usually the case, the further you get from the highway and the further you hike from the road.

There are a few mines in the region, primarily copper. The copper occurs in the sandstone and is called "red bed copper. " It is found with traces of malachite and azurite as well as fossil plants. Many of the mines are now closed and you should keep an eye out for the colorful associated oxides. Most are only surface stains and can't be used for more than garden stones; however, you might be lucky and find something thick enough to cut.

Parked along Forest Service Road 100

Pedernal Peak

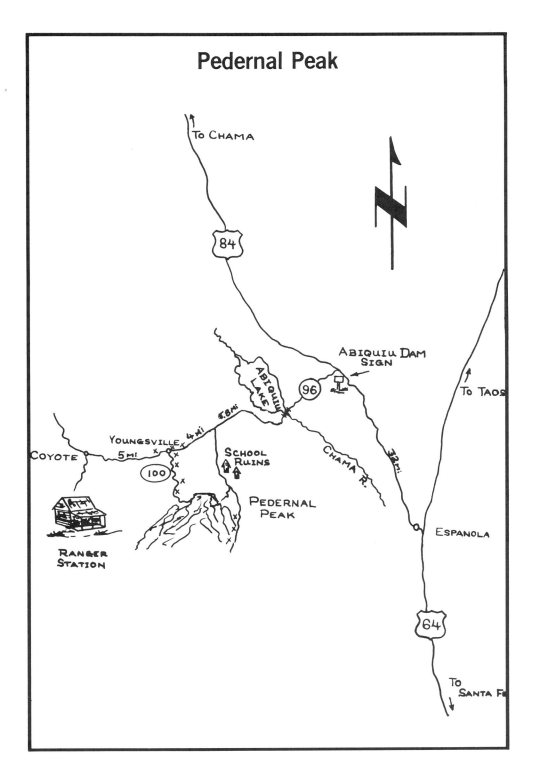

To Chama

84

Abiquiu Dam Sign

96

Abiquiu Lake

To Taos

4 mi

5.8 mi

Youngsville

Coyote

5 mi

100

School Ruins

Chama R.

32 mi

Pedernal Peak

Espanola

Ranger Station

64

To Santa Fe

41

BISTI BADLANDS

The Bisti Badlands south of Farmington have long been of interest to mineral collectors, photographers and paleontologists. They lie in what geologists refer to as the San Juan Basin, the remnants of an ancient, heavily forested swampland. Amongst the beautiful colored mounds, spires and cliffs, rockhounds can find petrified wood, common opal, coal, calcite, mud concretions, agate, jasper, quartz crystals, and even some fossils.

Start in Farmington and drive south thirty-three miles on Highway 371. The road is paved for the first 20 miles and then turns to dirt. It is well graded, though, and should present no problem to most vehicles. At the given mileage, you will be in the center of the Bisti Badlands. Park anywhere, and hike through this geological wonderland on either side of the road.

It doesn't take long to start finding something of interest. The wood, agate and jasper are scattered randomly all over the terrain in varying concentrations. The fossil shells are found in the blocky rocks that can be seen slipping down the sides of some of the mounds. Fossil animal remains are the most elusive of the Bisti treasures to locate and are usually discovered in gullies and ravines which have been freshly exposed by recent rainstorms. Look for quartz crystals in cavities of harder rocks while the calcite and common opal is more frequently encountered in the softer soil of the mounds. As you explore the badlands, be also on the lookout for coal, since contained in some of it are delicate fossil leaves and twigs. Some are so well preserved that they are great for display.

Bisti Badlands

Bisti

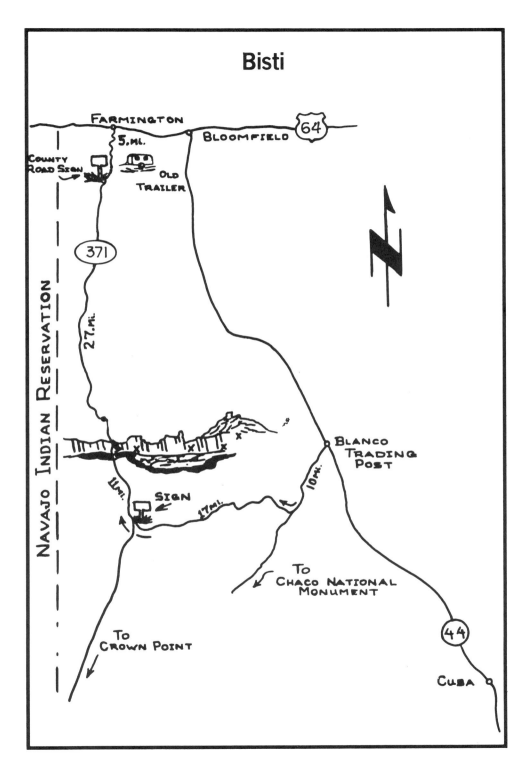

NORTH OF FARMINGTON

The region in the northwest corner of New Mexico known as the San Juan Basin has fascinated geologists and paleontologists for years. The colorful badlands of the basin have supplied scientists with a wealth of information about the prehistoric past and also offer rockhounds and fossil hunters countless opportunities to enjoy their hobby.

This particular site encompasses one small part of the San Juan Basin, is easy to get to and very productive. Go north on Highway 170 from Farmington exactly three miles. At that point, there are some ruts on the left which you should follow as they lead up the hill from the pavement. Along the way, agate and jasper can be seen beside the road, but continue to the flat area only four-tenths of a mile from the highway. From the flat, the tracks continue up a very steep hill onto a precarious ridge, and it is very dangerous to drive further.

Agate, jasper, opalite and rhyolite are plentiful on and around the road from the highway, in the parking area, along the ridge, and on most other upper regions. Petrified wood and fossils are primarily obtained in the badlands below and are much tougher to find.

The jasper occurs in a variety of solid shades of red and orange, while the agate encompasses a much wider range of hues and patterns. Beautiful black agate with brilliant red swirls and a good clean moss variety are among the favorite from the region.

The best overall collecting is on the upper areas and, surprisingly, not in the badlands themselves. When descending into the badlands to look for the petrified wood and fossils, be very careful, since the bank is steep and the soil loose, making it difficult to maintain your footing.

Collecting in the flat area below ridge

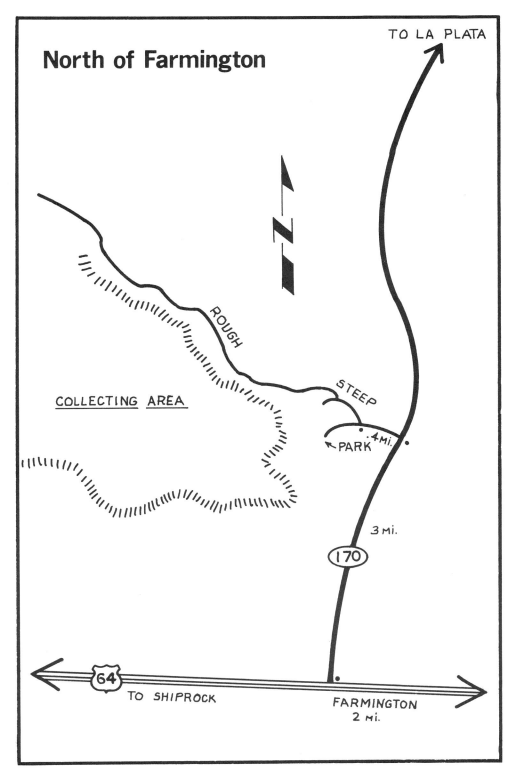

North of Farmington

TO LA PLATA

ROUGH

COLLECTING AREA

STEEP

.4 Mi.

←PARK

3 Mi.

(170)

(64)

TO SHIPROCK

FARMINGTON
2 Mi.

LA MADERA AREA

These three collecting sites afford rockhounds an opportunity to gather calcite crystals, mica books, geodes, pink feldspar, agate and jasper. To get to the first location, Site A, start in La Madera and take Highway 519 east across the bridge about one and one-half miles to the abandoned mica mine. It is difficult to spot the shaft from the highway, since it is partially hidden by brush; so pay particularly close attention as you approach the given mileage. Park well off the pavement and hike across the wash to the mine where you will find specimens of book mica, as well as vivid pink feldspar crystals.

Follow the rubble to the left of the shaft up the hill about fifteen yards, and you will see where others have been digging for calcite rhombs. Half way to the top is some rotten rock which contains terminated calcite crystals.

The turn to Site B is only a mile further along Highway 519. Just before rounding a hairpin, as the road climbs to the north, there is a set of ruts leading off to the right. Turn there, go only a few yards, park and hike down the bank to the opposite side. It is on the hill on the opposite bank where one can find a most interesting occurrence of calcite, a pseudomorph of ilmenite. The calcite has completely replaced the ilmenite and maintained its tabular crystallization. By digging into the soft soil, specimens of these mineralogical oddities can be found with patience and time. There are also cavities in the upper portions of the hill which contain terminated quartz and calcite crystals.

The final location, Site C, is not as productive as the other two but does offer an opportunity to collect additional minerals. Simply continue on Highway 519 one more mile. Then, primarily on the right side of the roadway extending into the canyon for at least another three-tenths of a mile, one can find agate and chalcedony scattered about.

At Site B

La Madera Area

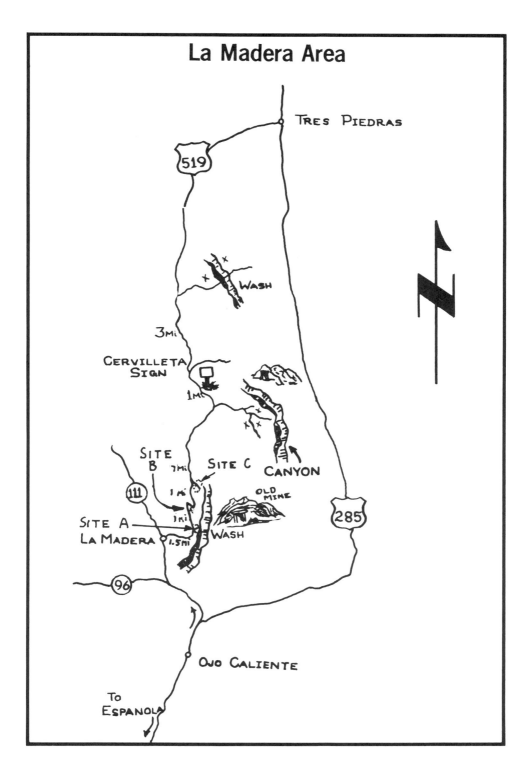

TRES PIEDRAS

519

WASH

3 MI

CERVILLETA
SIGN

1 MI

SITE
B   SITE C
7 MI   CANYON
111   1 MI
OLD
MINE
1 MI
285
SITE A
LA MADERA   1.5 MI   WASH

96

OJO CALIENTE

TO
ESPANOLA

47

DIXON AREA

Probably one of the best known pegmatite mines in all of New Mexico is the Harding Mine, located a short distance east of Dixon. This site is a mineral collector's dream come true, offering opportunities to find nice specimens of vivid purple lepidolite, rose muscovite, rhombohedral calcite, pink feldspar, blue apatite, green tourmaline and a host of other minerals. The mine is currently owned by the University of New Mexico, Department of Geology, and permission to collect must be secured before entering. It is necessary to sign a liability release which can be done on campus in Albuquerque or in Dixon at the home of either Alice Gilbert or Bernabe Griego. Directions to those homes are available at Labeo's Store, Dixon.

After securing permission, follow Highway 75 six and one-half miles east from town to a graded dirt road which intersects on the right. The road is difficult to spot because of shrubs and trees growing on the side. The mine is only six-tenths of a mile from the pavement, and the road is not too rough, being suitable for most vehicles if driven carefully.

There is a locked gate about 200 yards before the first dump, and you must park there and hike the rest of the way. Simply follow the road past the gate to the mine. The vivid purple lepidolite stands out brightly against the grey native rock and is easy to spot, even when standing. It takes a little more patience and closer inspection, however, to find the other minerals.

The optically clear rhombohedral calcite crystals are found by following a dim trail above the mine. The trail can be spotted as you approach the first dump from the parking area. The mine will be to the left, and the trail to the calcite is directly ahead, going up the hill, on the other side of the canyon.

Remember that entering mine shafts can be very dangerous, and the Harding is no exception. Lots of outstanding mineral specimens can be found on the dumps and in the exposed cuts, making entering the underground workings unnecessary.

Trail to the Harding Mine

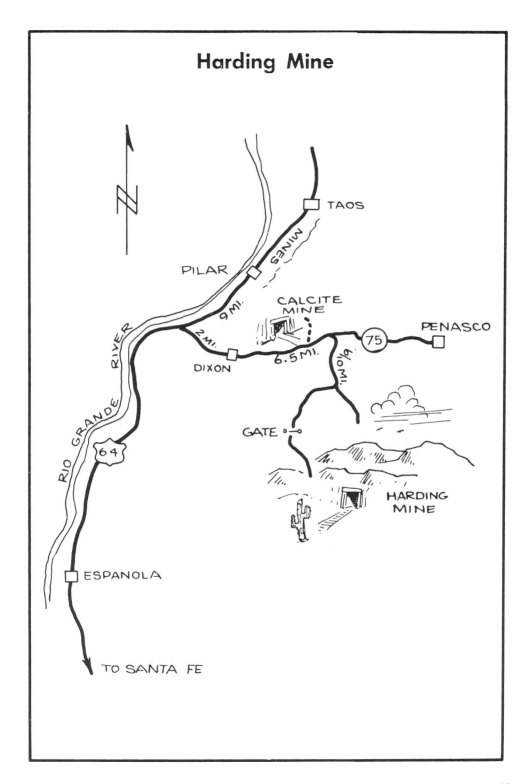

Harding Mine

TAOS

PILAR

MINES

CALCITE
MINE

PENASCO

9 MI.

RIO GRANDE RIVER

2 MI.

75

6.0 MI.

6.5 MI.

DIXON

64

GATE

HARDING
MINE

ESPANOLA

TO SANTA FE

49

*Chalcedony, opalite, and other specimens
from the Chavez Canyon area.*

*Smithsonite and other minerals collected
near the Kelly Mine*

*Petrified wood and agate found near
Albuquerque*

*Calcite pseudomorphs after ilmenite
surrounded by calcite crystals*

Chunks of garnet and stuarolite bearing schist - Taos Area

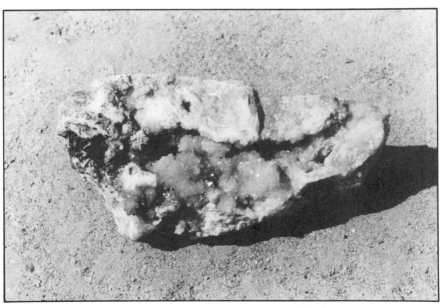

Crystal filled cavity in rock from the Blanchard claim near Bingham

TAOS AREA

There are many schist outcrops between the towns of Pilar and Taos, and embedded in some of that schist are garnets, book mica and staurolite crystals. Some of the staurolite displays twinning in the form of X shaped and right angle crosses called Fairy Crosses. These are highly prized among collectors throughout the world.

Most rockhounds believe the best of the deposits is four miles northeast from Pilar on Highway 68 just past milepost 33. At that point, a dirt road will be seen leading off to the right. Turn Here. Be advised that the ruts are very rough and washed out in places, being passable by only the most rugged of four-wheel drive units. If you don't think your vehicle can make the journey, park off the highway and hike.

About one mile further, there is a Forest Service gate and shortly beyond the road is completely washed out and totally impassable, even with four-wheel drive.

At that point, if not before, you must park and set out on foot. The trek isn't too bad but slightly uphill. The overhanging branches on both sides of the trail provide shade, and a little stream parallels the old roadbed.

About one-half mile past the gate you will start finding fragments of schist, some containing small garnet and staurolite crystals. Sizeable mica books can also be picked up, and a few might make good display pieces.

A little more than one mile beyond the Forest Service gate, the trail widens into a clearing, and just beyond this clearing is one of the prime staurolite bearing schist deposits. Break any suspect stone with a rock pick in hopes of exposing enclosed garnets, staurolite and mica. It takes lots of patience to find the crosses, but the effort is usually rewarded. Additional schist outcrops can be seen further up the canyon, and you might want to explore them.

Hiking along the trail to the primary collecting area

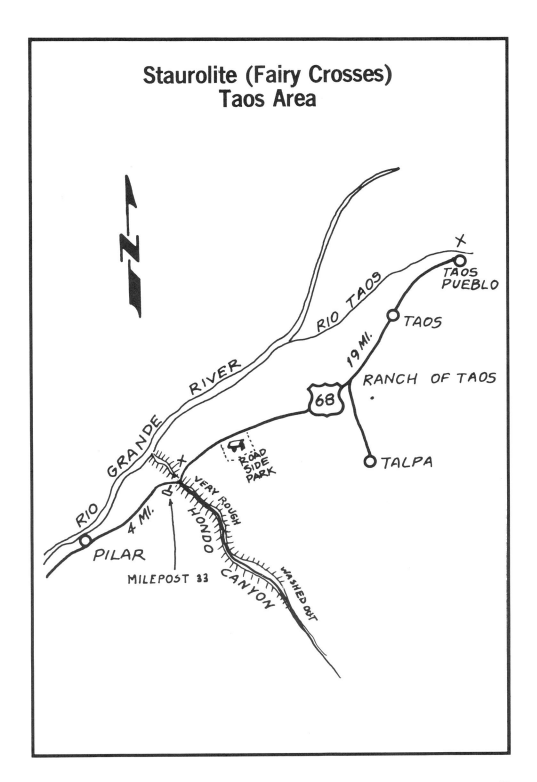

Staurolite (Fairy Crosses)
Taos Area

CIMARRON CANYON

At time of publication, this entire region was being reforested, thereby closing all roads and making it necessary to hike to the collecting site. In addition, no overnight camping was allowed, so plan accordingly if those regulations are still in force. If you choose to make the hike, it is about one and one-fourth miles and somewhat steep. Therefore, it is only recommended to those in good physical condition. Be sure to take along some drinking water and wear proper footwear.

To get there from Cimarron, go fifteen and seven-tenths miles to the Palisades Picnic Area, which is nestled at the base of a group of spectacular, sheer rock cliffs just north of the highway. This magnificent setting provides a good place to stop and get your equipment ready for the trek.

To get to the actual collecting area, drive eight-tenths of a mile further west, and you will see a road leading off to the north from the highway. Park well off the pavement and follow that road as it heads up the canyon. After having gone about one mile, there will be a log across the trail, and soon after that the road forks. Take either branch and proceed about three hundred yards to the jasper deposit.

There are lots of nice material to choose from, and many of the pieces have very weathered exteriors causing them to resemble chunks of colorful petrified wood. These unusual specimens often are desirable as is for display in mineral collections.

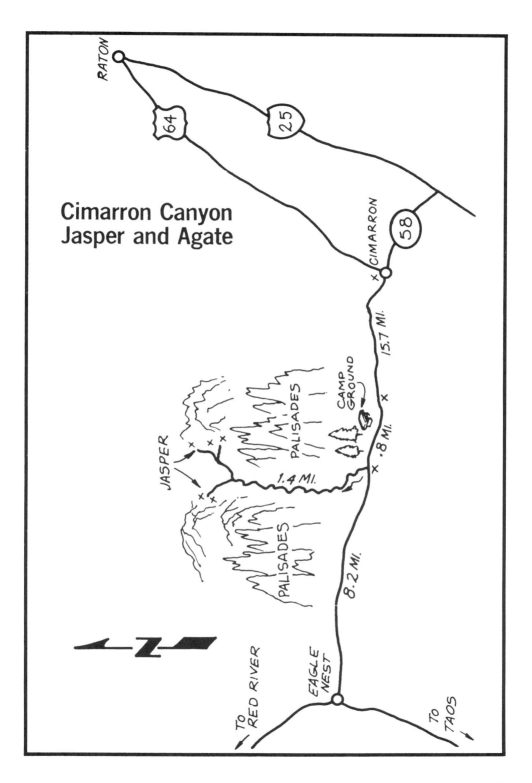

Cimarron Canyon
Jasper and Agate

RATON

64

25

CIMARRON

58

15.7 MI.

CAMP GROUND

PALISADES

.8 MI.

JASPER

1.4 MI.

PALISADES

8.2 MI.

TO RED RIVER

EAGLE NEST

TO TAOS

ROBERTS RANCH

This ranch is actually in Oklahoma, but since it is so close to New Mexico and such a good spot for rockhounds, it is mentioned in this book. At the ranch, Mr. and Mrs. Roberts have many samples of the various materials that can be found there, including petrified wood, algae, and cycads. From their yard, they can point to the Cimarron River where you find the algae, the mountain which contains the agatized limb sections, and to the other areas where they have picked up the cycads and a host of other collectables. A small fee is charged, but it is nominal when compared with what can be obtained.

The cycads are rare and difficult to find, but the other materials are fairly plentiful. It should be mentioned, however, that this location has long been known among rockhounds, and over the years lots of the surface material has been picked up. Nowadays, it does take some patience and exploring to find a good quantity of anything.

The ranch has a small grotto where an Easter pageant is held each year. If you are in the area at that time of the year, try to make plans to see it.

On the way to the ranch from Boise City, as shown on the map, you will pass the turn to Black Mesa State Park. There, you can find a good place to camp, or if you prefer, the Roberts do allow collectors to spend the night on their property.

There is a small store in Kenton about two miles from the ranch where you can obtain supplies. Of special interest to collectors, it also houses a small museum, displaying specimens of some of the minerals found in the area.

Agatized limb sections

Roberts Ranch

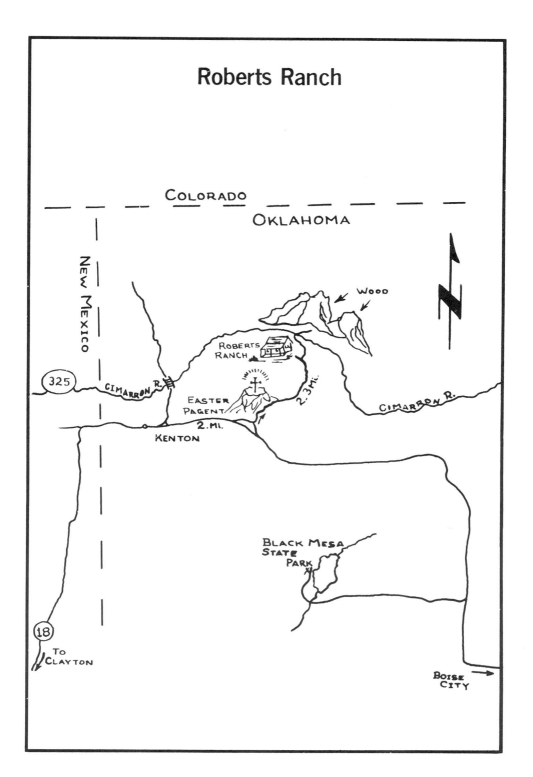

LAYTON RANCH

The Layton Ranch is very close to the tri-state marker, primarily in Colorado, but is included here since it is only a short distance across the state line. The minerals of greatest interest are petrified wood, algae, colorful rose agate and coprolite. A small fee is charged to collect on the property, based on how much and what is found. Mr. Howard Layton, the owner, will give you complete information when you visit.

This has long been a favorite collecting spot for rockhound clubs. Subsequently over the years, much of the surface material has been removed. Most collecting is done in the washes, and due to the scarcity of surface collectibles, you will probably have the best luck after a heavy rainstorm or following a wet season. The rains often expose otherwise buried material, but it is getting tougher and tougher to find large amounts of anything.

When you visit, Mr. Layton will show you specimens of what can be found on his extensive ranch and will direct you to Carrizozo Creek and its surrounding banks. Carrizozo Creek is considered to be the most prolific of his collecting areas, and it is fun to wade in the clear, cool, shallow water while looking for the algae, wood, rose agate and coprolite. Pay particularly close attention to gravel bars, since they usually contain much better concentrations of rocks and minerals than random spots along the river.

The coprolite is, by far, the most scarce of the Layton Ranch minerals, but if you are lucky and/or persistent enough, you may be able to find some. The other materials are more plentiful.

In addition to the Layton Ranch, there are two other supplementary collecting opportunities in the region. The first is a small hill near the tri-state marker, as shown on the map, which contains lots of nice rose agate. The second, also shown on the map, is south of the ranch on the Carrizozo River. Simply hunt the creek bed, as you did before, for similar minerals. Once again be sure to pay close attention to the gravel bars and banks.

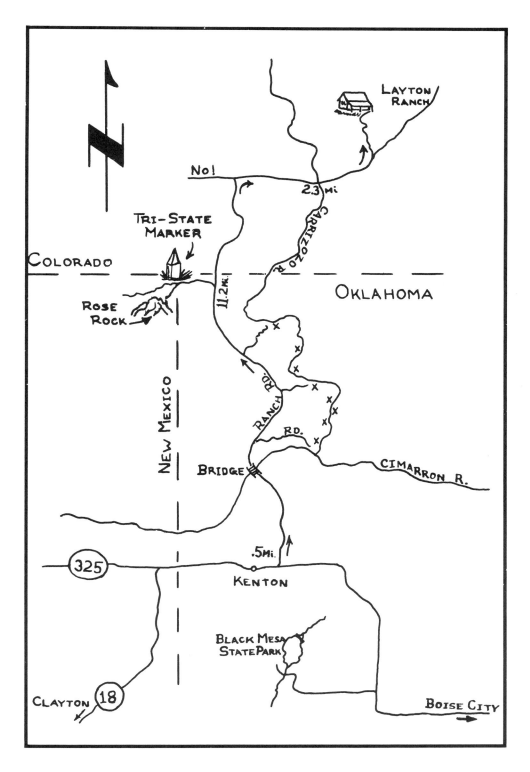

TERRERO MINE

The Terrero Mine, located about eighteen miles east of Santa Fe, has been a favorite collecting spot among rockhounds for many years. Because of that fame, however, much of the surface material on the extensive dumps has been thoroughly worked over. Nonetheless, this still can be a most productive site if you are willing to work and spend some time.

To reach the dumps, take Highway 50 north from Highway 85 about four miles to Pecos. From there, continue north on Highway 63, as shown on the map, about fourteen miles to the Terrero Store. The road is not paved past Pecos, but it is well maintained and should not present a problem to most vehicles.

From the store, it is necessary to go right and up the hill another mile. You will see the extensive dumps as you approach the mileages. They continue almost all the way to Willow Creek, where there is a nice campground if you plan to spend the night.

The list of what can be found here is extensive, but the minerals of primary interest are pyrite, mica, actinolite, garnet, tourmaline, lepidolite and bornite. To find thebest material, it will probably be necessary to use a pick and shovel to dig a few feet down. Splitting suspect stones to expose fresh surfaces will help greatly in the evaluation of potential specimens. In addition, it is helpful to have a bucket full of water handy to clean your finds. That will also greatly assist in the process of identification.

Terrero Mine (Pecos Canyon)
Mineral Specimens

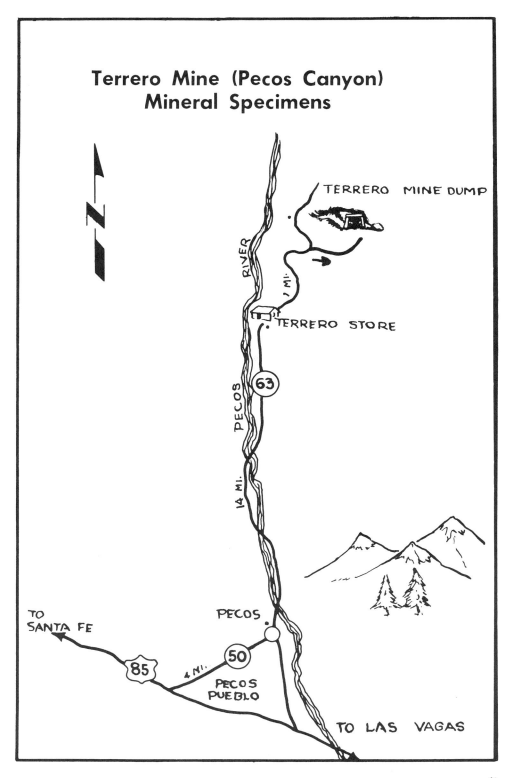

TERRERO MINE DUMP

RIVER

1 MI.

TERRERO STORE

PECOS

63

14 MI.

TO
SANTA FE

85

PECOS

50

4 MI.

PECOS
PUEBLO

TO LAS VAGAS

RED CLOUD MINE

The Red Cloud Mine, which actually consists of two distinctly different prospects across the road from each other, boasts a long list of minerals available to collectors. Most specimens, however, are quite small, being of primary interest only to micromounters.

To visit the renown Red Cloud Mine, take Forest Service Road #144 west from where it intersects Highway 54, about thirty-four miles north of Carrizozo and just south of Corona. Take this well graded dirt road three and one-tenth miles, turn left onto Forest Service Road #104 and proceed another two and seven-tenths miles. At that point, turn right and continue three and four-tenths more miles to the mine site. The road is fairly good all the way and most vehicles should have no trouble getting there.

Fluorite was mined on the right and copper on the left. At the former, dedicated collectors can obtain flattened, hexagonal crystals of waxy yellow bastnaesite and small yellow-green balls of agardite. In addition, colorful purple fluorite cubes and white barite are also encountered. A very interesting mineral occurrence at the Red Cloud fluorite mine is the goethite and hematite pseudomorphs of pyrite which are also fairly plentiful.

The minerals found at the copper mine are more colorful and include blue chrysocolla, grey cerussite, pyramidal black mottramite, reddish brown vanadinite, bright orange wulfenite, botryoidal green conichalcite, and bright yellow mimetite.

To find the above listed minerals it is usually necessary to split suspect stones in hope of revealing otherwise concealed mineral-filled cavities or fissures. Do not enter any of the shafts at either mine, since most of the native rock is quite rotten. Work only in the dumps or on the quarry walls.

Searching the dumps at the Red Cloud Mine

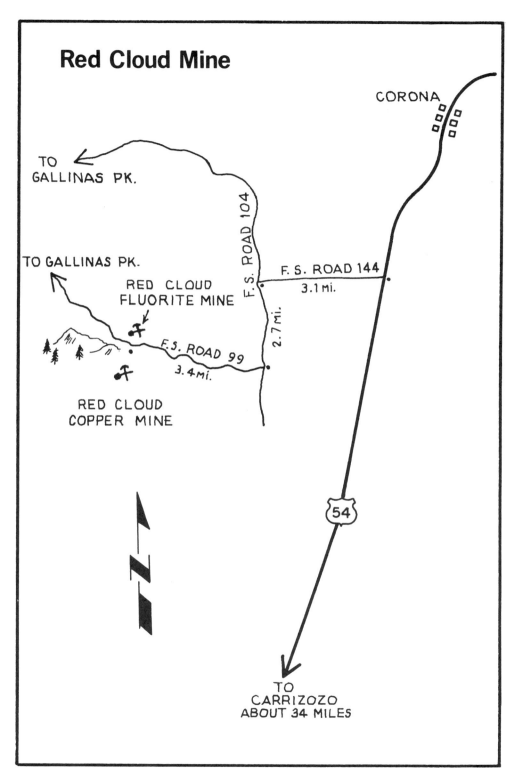

Red Cloud Mine

CORONA

TO
GALLINAS PK.

F. S. ROAD 104

TO GALLINAS PK.

RED CLOUD
FLUORITE MINE

F. S. ROAD 144
3.1 mi.

2.7 Mi.

F. S. ROAD 99
3.4 mi.

RED CLOUD
COPPER MINE

54

TO
CARRIZOZO
ABOUT 34 MILES

HANSONBURG MINING AREA

In the Oscura Mountains of Socorro County near the small town of Bingham are the famous Blanchard claims. Over the years, people have come from all over the world to collect there, and few ever went home disappointed.

The Hansonburg (Blanchard) mining area is currently open only to surface collecting on a supervised fee basis, and advanced arrangements to visit are usually needed. The effort to make these arrangements, however, is frequently rewarded with outstanding specimens of hard to find minerals such as atacamite, brochantite, celestite, cerussite, cyanotrichite, murdochite, plattnerite and spangolite. In addition, more common minerals can also be found, including barite, dolomite, fluorite, limonite, malachite, azurite, galena and quartz.

The claims are currently operated by Sam and Vera Jones. To make advance reservations or to obtain more information, contact them at the Blanchard Rock Shop, Highway 380, Bingham, New Mexico 87815. If you do visit, allow enough time to see their shop. There, you will be able to examine specimens of minerals found on the claims, and have a better idea as to what you will be looking for.

Be sure to take a rock pick and a chisel to split and/or trim suspect stones. It doesn't take much effort to find samples of minerals most collectors would never have an opportunity to find elsewhere. Even the more common minerals are frequently spectacular with the cubic galena and fluorite, occurring on a bed of quartz crystals, being one of the most prized.

Cubic galena on bed of quartz crystals - Hansonburg

Hansonburg Mining Area

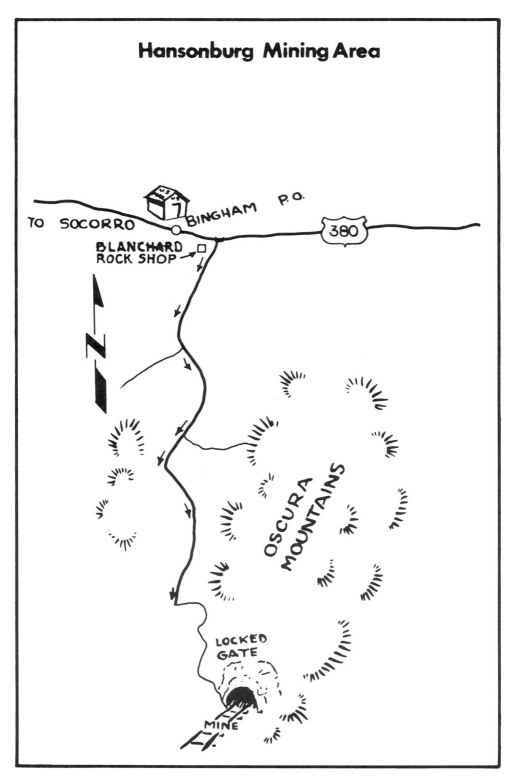

*Calcite specimens from Harding Mine
near Dixon*

*Petrified wood and agate found in
Los Lunas area*

Volcanic bombs from the Kilbourne Hole

Radium Springs banded marble chunks

Mineral specimens from Pinos Altos

Pyrolusite dendrites on the white native rock - Rincon Quarry

DEMING AREA - AGATE

The area around Deming has long been known for its colorful agate. One of the best concentrations of this often spectacular cutting material can be obtained a few miles south of town on a private claim owned by Mrs. Eddie Lindberg. The claim is operated as a fee site, but the charge of fifteen cents per pound is nearly nothing when you consider the quality of what can be found. Before following the directions given on the accompanying map, be sure to first check with Mrs. Lindberg at the Deming Agate Shop on the northeast side of town. There you will be able to get last minute information about the "Big Diggins" site, as well as have an opportunity to see some specimens that have come from there.

To get to the claim, head south from Deming on Highway 11 eight and one half miles to the turn at Sunshine School. Go right and continue four miles to the stop sign; then turn left and proceed six more miles. At that point, you must again turn right and continue another six miles on a somewhat rough dirt road. At the given mileage, just before the fence, as shown on the map, you will be in the center of the agate claim area.

Since these claims have not been operated on a commercial basis for many years, most of the surface material has been thoroughly picked over, making it necessary to do some digging to find the best the site has to offer.

It is impossible to describe the beauty and variety of what can be found here. The colors range from vivid yellow, orange and red through deep blue and black. The inclusion patterns seem infinite, with moss, lace and swirled motifs the most common. Much is solid and will work up into some of the most beautiful and colorfully patterned polished pieces imaginable.

Be sure to return to the Deming Agate Shop to weigh and pay for what you decide to keep.

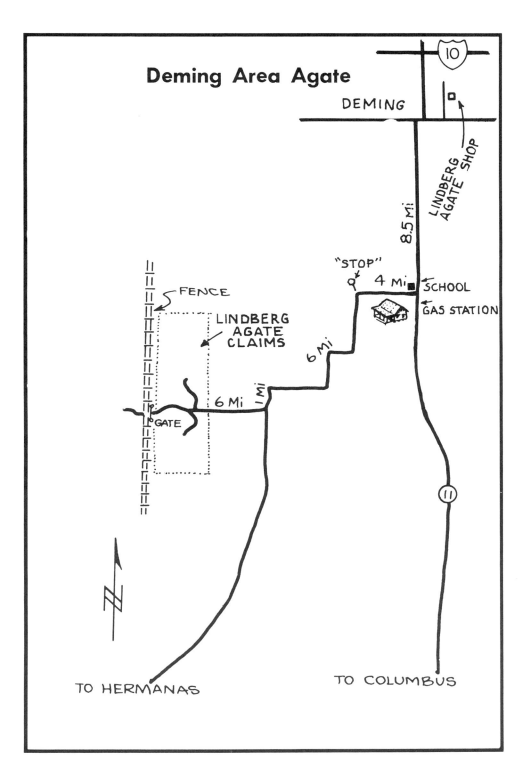

Deming Area Agate

DEMING AREA - NODULE BEDS

One of the best places in all of New Mexico to find top quality agate nodules is about thirty-eight miles southwest of Deming on the Lindberg nodule claims. The beautiful spheroids found in this area can be either solid or hollow with the latter occasionally being filled with beautiful purple amethyst crystals. The solids display a seemingly infinite variety of agate interiors, the most common being either banded or picture material, often filled with interesting inclusions. In addition, some of the nodules contain regions of blue-white opal, which in combination with the surrounding agate, are especially beautiful when cut and polished.

A 1951 article in National Geographic Magazine featured this location, and its popularity is well justified. If you would like to visit, it is necessary to first stop at the Lindberg Agate Shop in Deming to get final instructions from Mrs. Lindberg and arrange payment of fifteen cents per pound for what you decide to keep. The shop is easy to locate, being a large building with "Agate Shop" painted on the side just north of the main road on the east side of town.

At the shop you will not only be able to get up to date information about road conditions to the site, but can also see samples of nodules which were found there. A few years ago, the ground was regularly dug up with a tractor to aid visiting rockhounds, but since the death of Eddie Lindberg, that service had to be discontinued. It is now necessary to dig for the nodules, and it is tough work. Be sure to take a pick and shovel as well as plenty to drink. The beautiful nodules and agate that can be found there, however, will more than compensate for that extra effort.

Beautiful solid nodules from Deming area

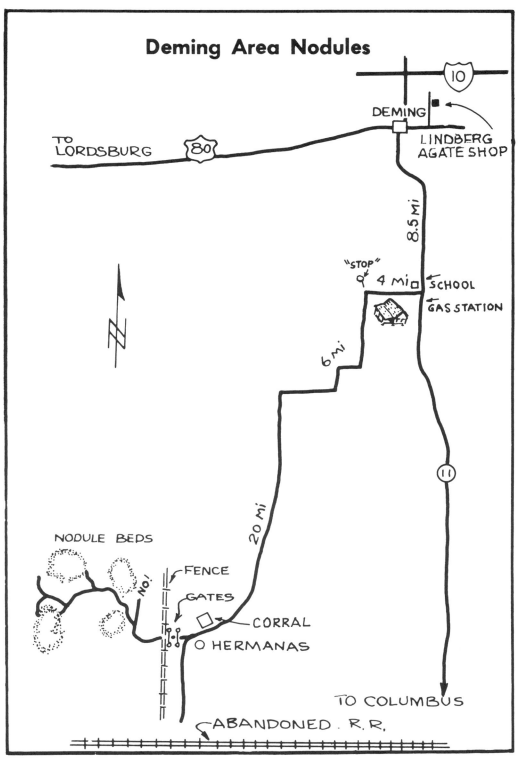

Deming Area Nodules

KINGSTON AREA

The area near Hillsboro offers the rockhound two interesting collecting locations. To get to the first, head west from town on Highway 90 about two and three-tenths miles to a large metal bridge crossing Petaca Creek. Stop at a convenient spot well off the pavement, and search east of the bridge for interesting spotted rhyolite. In addition to the unusual brown material filled with white dots, colorful banded rhyolite can also be found. Most does not take a high polish, however, but can be used to produce nice spheres, bookends and large cabochons. Additional similar material can be found at the next bridge, which is two and seven-tenths miles further west. Be careful if walking on the highway, since drivers are not looking for pedestrians or parked cars on this narrow stretch of road.

The second site, which features some unusual forms of crystalline quartz, is about nine more miles to the small town of Kingston. Bear right in town, leaving the highway, and continue two-tenths of a mile. At that point, take the dirt road heading north another seven-tenths of a mile to a gate and stop. The crystals are found on the hill to the right in seams as well as loose in the soil. The easiest way to recover the crystals is to sift through the dirt on hands and knees with a small garden rake or trowel hoping to uncover some hidden specimens. This is tedious work but usually pays. The most productive method of working this area, however, is to split the large boulders found throughout the hillside in hopes of exposing a crystal bearing seam or cavity. This requires some tough work and heavy tools, such as a sledge hammer, gads and chisels.

Most of the crystals found here display single terminations, but some are doubly terminated, prize scepters or clusters of all types.

Parked at the rhyolite deposit

Quartz Crystals

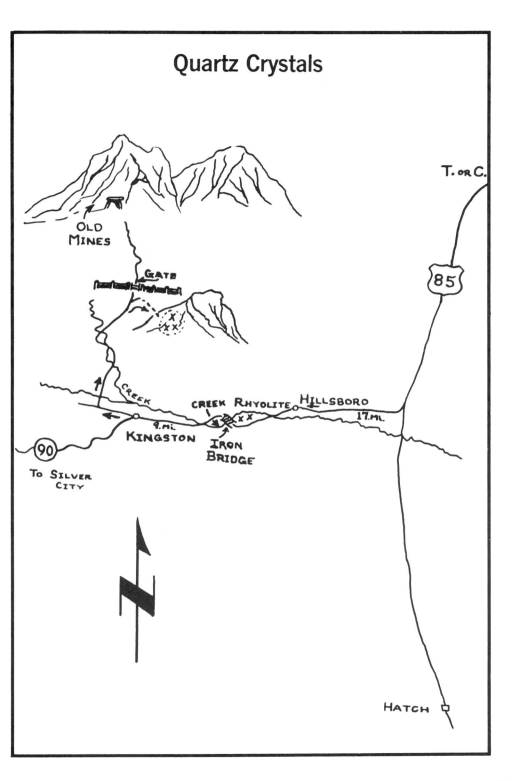

LAKE VALLEY

The ghost town of Lake Valley is situated in the center of a well known mineral collecting region. To get there, take Highway 27 ten and seven-tenths miles south from Hillsboro to where an outcrop will be seen on the left side of the road. It is on and around this little hill and in the flatlands on both sides of the dirt road just north of it where nice agate can be found.

There is a large quantity of material, especially near the highway. It is often colorful, occurring in tones of red, pink, green and black, some displaying an interesting patchwork design. Other pieces are often filled with swirls and bands of color making them highly desirable. Much of what is found here is fractured, however; so be careful to take sufficient time to sort out the good from the bad. In addition to the agate, small calcite crystals can be found. They are located in the terrain on both sides of the dirt road, as well as being scattered next to the paved highway. The crystals are not too plentiful but worth looking for.

To get to Lake Valley, continue south five and four-tenths miles. Be advised that at the time of publication the entire ghost town and its associated dumps were closed to collectors and are mentioned only if that status should change. There is a full time caretaker on duty at Lake Valley, and he will let you know if any of the dumps are open at the time of your visit. If they are, it is possible to find good specimens of pyrolusite, magnetite and psilomelane, in addition to nice fossils. Even if you are not allowed to do any collecting at Lake Valley, it is still interesting to see, from a distance, the remnants of this once bustling silver town.

Looking for calcite crystals near road

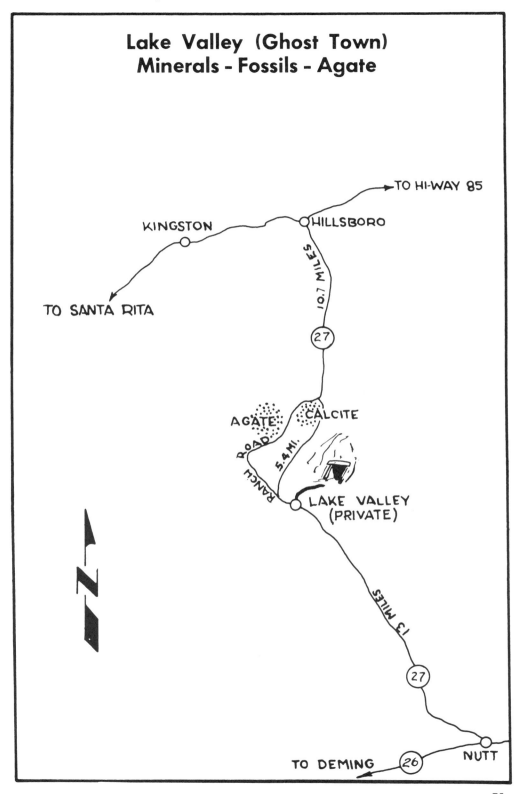

Lake Valley (Ghost Town)
Minerals - Fossils - Agate

TO HI-WAY 85

KINGSTON

HILLSBORO

TO SANTA RITA

10.7 MILES

27

AGATE

CALCITE

RANCH ROAD

5.4 MI.

LAKE VALLEY
(PRIVATE)

13 MILES

27

TO DEMING

26

NUTT

75

TRUTH OR CONSEQUENCES

Candy rock is the name local collectors have given to the banded rhyolite found a short distance west of Truth or Consequences. This elegant material is very popular for use in building walls and fences in this part of New Mexico, as well as finding applications in smaller lapidary projects such as spheres, bookends and cabochons. The color ranges from dark reddish brown to a cream color and frequently displays highly contrasting bands.

To get to the primary collecting area, which is only a very short distance from the center of Truth or Consequences, go north on Date Street to 9th Street and turn west following 9th Street to Interstate 25. At the freeway, there is a "tunnel" which locals call "The Whistle" leading under the highway. Small cars can drive through to the ranch road on the other side, but if you have a camper or large vehicle, you may have to park and walk the rest of the way. Be certain before driving through that your vehicle will fit.

Whether you drive or hike, follow the ranch road west approximately one mile to the south side of the mountain with the large antenna on top. When at the appropriate mileage, it will be easy to see where others have been digging before. This marks the collecting site.

Small chunks can be found lying around, and it only takes a short time to gather a good amount of cabochon size material. Just be sure to take time to gather only that which displays the finest banding and best color contrast. If you want to obtain larger pieces, it is necessary to use a sledge hammer, pry bar and other hard rock equipment to directly attack the deposit in its place in the mountain.

Be sure to leave all gates the way you find them, since there may be cattle grazing here.

Banded rhyolite or 'candy rock'

Candy Rock

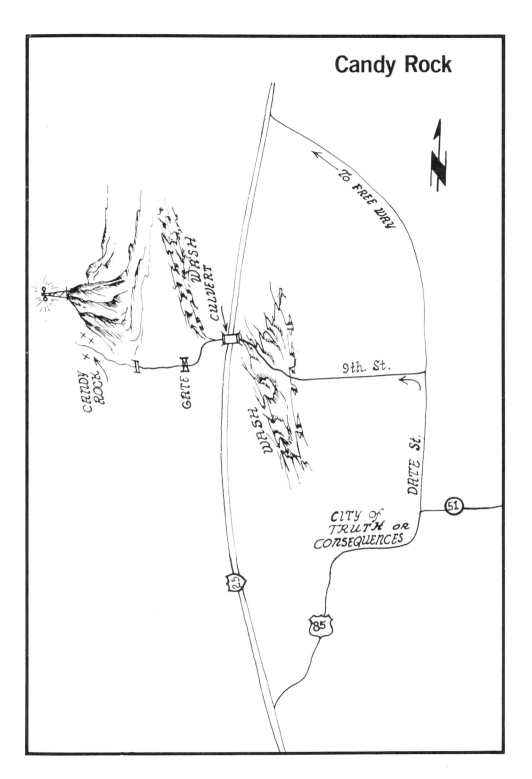

COOKS PEAK

Small gem-quality pieces of carnelian agate can be found a short distance northeast of Deming. The collecting area covers hundreds of acres traversed by numerous roads and trails.

Take Highway 180 north about one and two-tenths miles from the center of town to Highway 26. Go east on Highway 26 approximately four and nine-tenths miles to Krol's Ranch Road where there is a rock shop at the intersection. Proceed north on this well graded dirt road past the shop and through the hills approximately six more miles. At that point you will be at the southwestern edge of the extensive collecting area. Park on any of the side roads past the given mileage. The collecting extends down the main road for at least twenty more miles all the way to the base of prominent Cooks Peak.

The beautiful red and orange carnelian is usually small, ranging from one-fourth to one inch across, but occasionally a lucky rockhound is fortunate enough to find a much larger piece. Even the tiny pebbles, however, make nice tumbled stones and can be used in many jewelry applications.

One can also find bright red agate, filled with vivid white lace-like patterns, as well as interesting brecciated jasper, quality picture agate, and lots of common agate which occurs in a variety of colors, including clear, black and white, often filled with a host of different and sometimes colorful inclusions.

The agate is not concentrated in any particular location, but is randomly scattered from hill to hill throughout the region. Some areas are practically barren while other places are virtually covered with material. The hunting is usually best after a rain or wind blow as previously buried agate is exposed by storms. If you have patience and are willing to spend some time and do some walking, you will surely find enough to make the trip worthwhile.

Carneliain agate and red jasper

Carnelian Agate

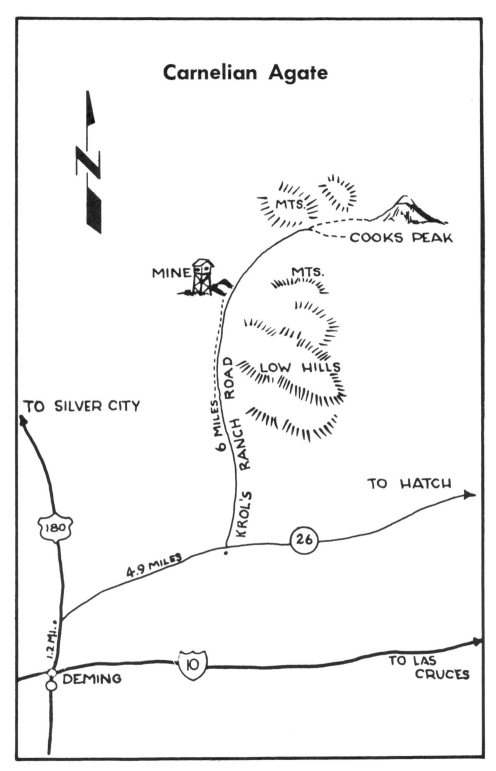

RADIUM SPRINGS

In 1949, while rockhounding near their home in Radium Springs, O. W. Preece and his wife discovered surface outcrops of very colorful banded marble. They filed thirty-five claims on the deposit a few years later.

The marble is called "rainbow marble" because of the incredible variety of different color bands that are displayed in choice pieces. Nearly every color of the spectrum can be found in this gorgeous material, except shades of blue. Even an emerald variety occurs as do browns, reds, yellows, golds, oranges, black, white and many shades of pink. The material is relatively soft which makes it easy to cut and polish.

A charge is made to collect there and you must stop at the post office in Radium Springs to pay the fee before heading in. At time of publication, the charge was $3.00 per person, per day, with a thirty pound limit.

Drive north one and one-half to two miles watching on the left for the dirt road heading up the hillside. Proceed another four and three-tenths miles to a fork and take either branch to one of the many marble quarries on the hill. The road isn't too bad but is rough in a few places.

Small chunks can be found scattered all over the quarries, but if you want large pieces, it will be necessary to remove them with hard rock equipment such as gads, chisels and sledge hammers.

Agate bearing rhyolite nodules, some displaying nice bands of white, grey, red and orange, can be found near the water tank. It is necessary to do some digging, however, for the best nodules.

Banded marble near Radium Springs Quarry

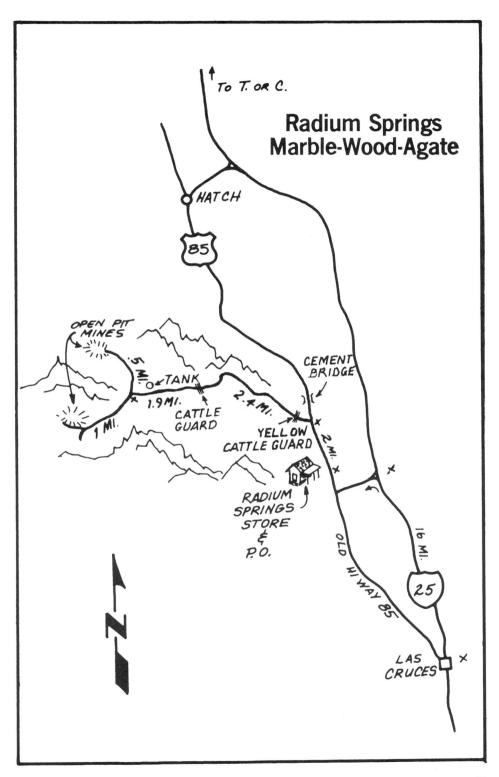

Radium Springs
Marble-Wood-Agate

TO T. OR C.

HATCH

85

OPEN PIT MINES

.5 MI.
TANK
1.9 MI.
CATTLE GUARD

1 MI.

2.4 MI.

CEMENT BRIDGE

YELLOW CATTLE GUARD

2 MI.

RADIUM SPRINGS STORE & P.O.

16 MI.

OLD HI WAY 85

25

LAS CRUCES

MULE CREEK AREA

Apache tears are always in demand by the rockhound. This location provides the collector with an opportunity to find lots of small, generally opaque specimens which can be made into beautiful cabochons or faceted stones.

To get there, take Highway 180 northwest from Silver City about forty seven miles to Highway 78. At that point, you should turn west and head for Mule Creek about nine miles further. From Mule Creek, continue another five and two-tenths miles to the site, the center of which is the state line. There is a turnout on the south side of the road, and the Apache tears can be found scattered everywhere for quite a distance on both sides of the highway throughout the pine trees. Be certain to park well off the pavement and be cautious when crossing the highway.

This collecting site is particularly nice during the summer months when you can roam through the shade of the tall pine trees as you look for the beautiful black gemstones. The best time of the day to search for Apache tears is in the early morning or the late afternoon. Keep the sun to your back and the glass tears will catch the sun's rays and sparkle as you walk along.

While in the area, you might want to take a side trip south of Mule Creek to the old Carlisle Mining area where amethyst crystals and other minerals of interest can be found on many of the dumps. Before collecting at any of these potentially productive mines, however, be sure to ascertain which are abandoned and which are not.

Further information can probably be obtained in Mule Creek or Silver City.

Parked in the turnout near collecting area

Apache Tears (Obsidian)

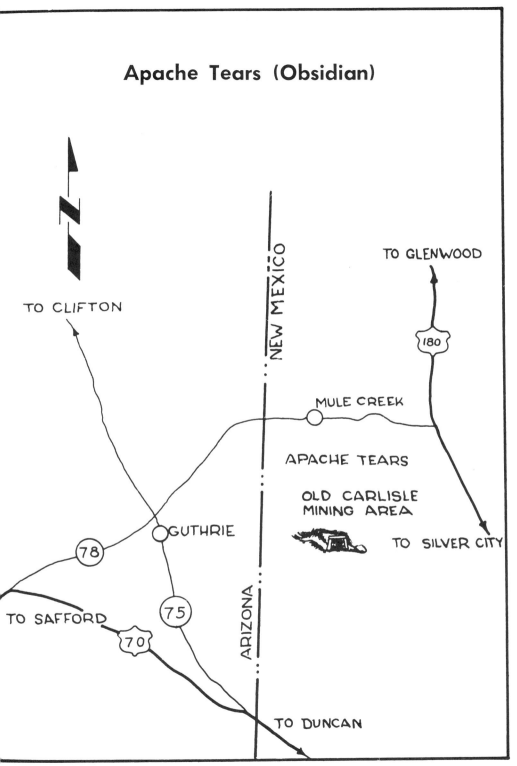

PUEBLO CAMPGROUND

Small gem quality bytownite, labradorite and hypersthene can be found near the Pueblo Campground in the far west part of the state. Take Highway 180 about 21 miles north from Alma to Forest Service Road 232. Turn there and proceed west six and one-tenth miles to the Pueblo Campground. The collecting site is directly across the road from the campground and is reached by entering through a gate at the head of the W. S. Mountain Trail. You can park at the trailhead or, if you plan to spend the night, at the campground.

The gemstones were formed in the gas cavities of the reddish rhyolite which can be seen on the slopes of the hill directly ahead and to the left as you start along the W. S. Mountain Trail. The most productive of the rhyolite seems to be that which is rust red in color and sprinkled with white specks. Once you find the first such piece, you will know exactly what to look for. Crack any suspect rhyolite in hopes of exposing fresh crystals, but be careful that they don't fall out and become lost. Some of the gems measure as much as an inch across, but most are far smaller.

In addition to finding them in gas cavities of the native rhyolite, loose crystals can also be found lying on the ground throughout the region, especially in the lower areas. These loose specimens are particularly easy to find after a recent rainstorm, since they will have been freshly cleaned and will sparkle in the sunlight.

Much of what can be found here is of faceting quality, and if the crystals are sizeable enough, beautiful and sometimes valuable pieces can be made from them.

Occasionally white agate, which takes a nice polish, can be found in washes and other low-lying areas.

Samples of the rust-red rhyolite with white specks

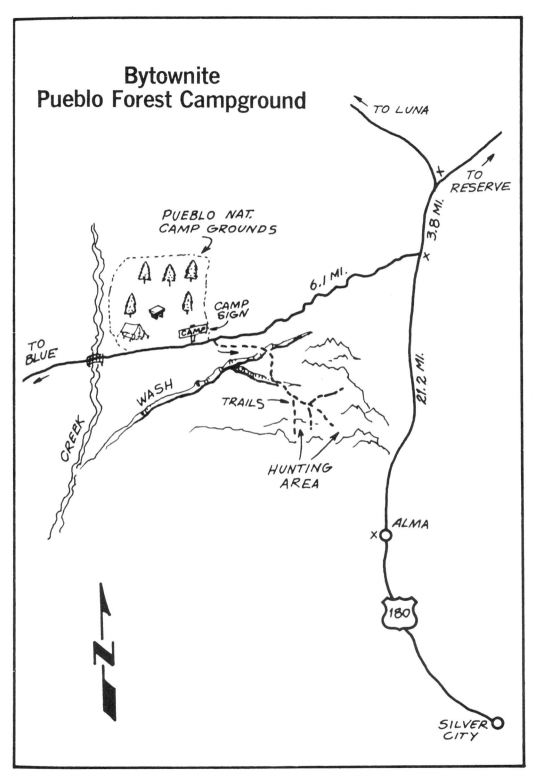

Bytownite
Pueblo Forest Campground

TO LUNA

TO RESERVE

3.8 MI.

PUEBLO NAT.
CAMP GROUNDS

6.1 MI.

CAMP
SIGN

CAMP

TO
BLUE

WASH

21.2 MI.

CREEK

TRAILS

HUNTING
AREA

ALMA

180

SILVER
CITY

N

BEAR MOUNTAIN FOSSILS

Crinoid stems, brachiopods, horn coral and a host of other well preserved fossils can be found a short distance northwest of Silver City. This has long been a favorite fossil collecting spot for New Mexico residents and is accessible by most vehicles.

To reach this highly productive region, take Alabama Street north from where it intersects Highway 180 on the east side of town. It changes to Cottage San Road (Forest Service Road #853) only a short distance after leaving the highway, and the pavement ends about three miles further.

The most accessible of the collecting areas, Site C, is encountered two and one-half miles past pavement's end. At that point, there are some ruts leading off to the right, and the hills on both sides of those ruts extending for at least one-quarter of a mile are filled with well preserved fossils.

Split any of the limestone and the freshly exposed surface will often be covered with crinoid stems and small horn coral. Occasionally other ancient life forms will also be discovered. It doesn't take much effort to find outstanding specimens in and out of matrix.

The limestone outcrop, Site B, can be seen from the road four-tenths of a mile further along. However, you must hike through the brush up the side of the mountain to reach the outcrop. As before, simply split the tough limestone in hope of exposing the fossils.

Site A is similar to Site B, since you must hike up the hill through the brush and trees to get to the easily spotted outcrop. The fossils at Sites A and B tend to be smaller and less plentiful than those at Site C.

Throughout this region, you will see many less accessible limestone outcrops which may offer further potential.

Fossil-filled hill at Site C

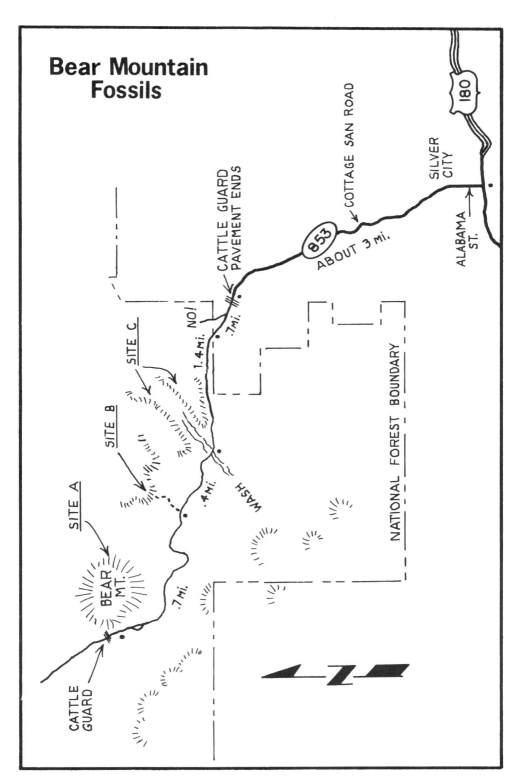

Bear Mountain
Fossils

SITE A

SITE B

SITE C

CATTLE GUARD
PAVEMENT ENDS

NO!

1.4 Mi.

.7 Mi.

.4 Mi.

WASH

.7 Mi.

CATTLE
GUARD

BEAR
MT.

853

ABOUT 3 Mi.

COTTAGE SAN ROAD

SILVER
CITY

ALABAMA
ST.

180

NATIONAL FOREST BOUNDARY

PINOS ALTOS

This location boasts a variety of collectibles, but nothing is particularly plentiful. It takes some patient hiking to find acceptable quantities of anything, but if visited in conjunction with a trip to the Gila Cliff Dwellings or as part of a leisurely drive through the mountains, it makes a great place to stop.

The site is about twenty-seven miles north of Silver City. Take Highway 15 north, as shown on the map, to where Highway 35 intersects. There is a turnout exactly two miles further, and you should park there. This is the center of the collecting area and a good starting point.

Walk along the creekbed east of the road for quite a distance and through the surrounding hills on both sides of the highway. If you climb on the slopes, be very careful, since the soil is soft and the loose rock sometimes makes it difficult to maintain stable footing.

The region boasts interesting bird's-eye and banded rhyolite, especially on the west side of the road, lots of common white agate, beautiful red agate, brown and black polka-dot jasper and occasional pieces of botryoidal hematite. There is some fossil bearing limestone scattered around, sometimes filled with crinoid stems and/or miscellaneous shells.

Much of the agate and jasper is weathered on the surface, making it look very ordinary, almost like the abundant rhyolite of the region. Be sure to split any suspicious stones to ascertain their true nature. You may have a very pleasant surprise!

Collecting along creek bed

Pinos Altos

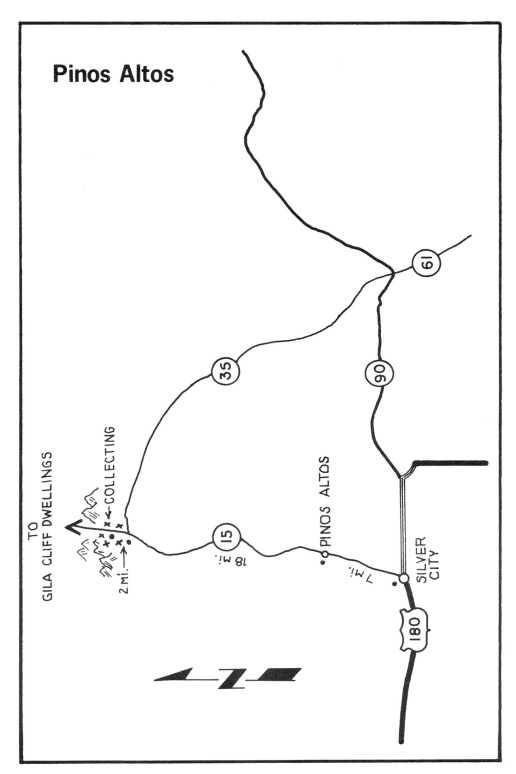

ROCKHOUND STATE PARK

The state of New Mexico has set aside approximately 250 acres near Deming for the exclusive use of rockhounds. This unique site is called Rockhound State Park, and within its boundaries one can find an incredible variety of minerals. A campground is situated in the low saddle of the Little Florida Mountains, and you must hike from there to the various mineral deposits. The terrain is somewhat rough, but there are many trails crisscrossing through the region to help make the exploration easier. However, it still seems that the best material is always found in the hardest to reach places.

Proceed east on the old highway five and one-half to six miles from the Lindberg Agate shop. Turn south and continue another five and nine-tenths miles, following the signs to the well marked State supported collecting area. There are campsites, restrooms and even showers, making it a welcome refuge, especially when on a lengthy trip. With nightfall, the campground affords a spectacular view of Deming in the valley below. At time of publication, there is a $3.00 charge for overnight camping, or $4.00 if you desire electricity.

Near the park's entrance, there is a display of the many minerals that can be found there, including thunder eggs, rhyolite, jasper nodules, geodes, yellow jasper which is often filled with beautiful red and black stringers, orange jasper, and black, brown and grey perlite. Quartz crystals, pink jasper, jasp-agate, variegated jasper and a deep chocolate brown jasper can also be found. It is no wonder why it was decided to dedicate this particular spot as Rockhound State Park. The road is paved all the way to the campground, and any type vehicle should be able to make the journey.

The best material seems to come from the higher areas above the campground in the little canyons, and some exploration is necessary to find the particular minerals you are seeking.

Another area to search is reached by following the road through the low saddle to South Canyon. Some spectacular large geodes have been found on this side trip.

Geodes and Agate Indian Artifacts
Spanish Stirrup Ranch

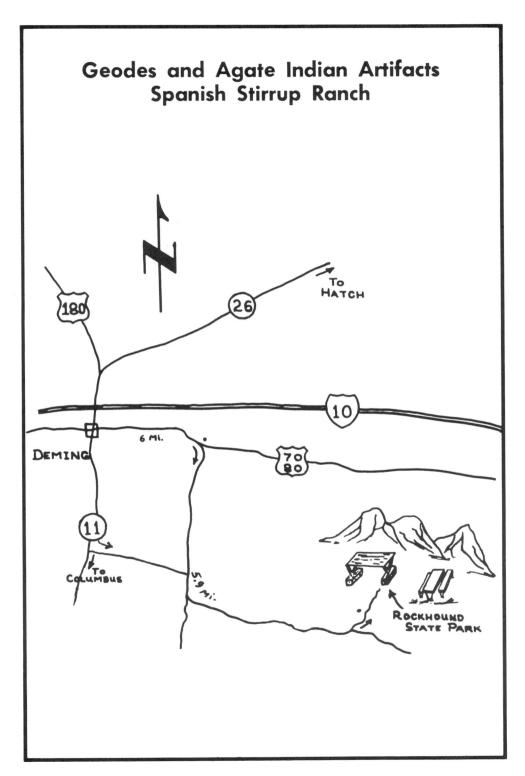

SHAKESPEAR GHOST TOWN

The region south of Lordsburg is of special interest to mineral collectors and those who enjoy ghost towns. The hills are filled with mine dumps and in the center of the area is the old town of Shakespear, which in recent years has had the exteriors of a few buildings restored. The town is privately owned, but every second Sunday of the month guided tours are conducted for the nominal charge of $2.00 per person.

The General Merchandise Store has been fully restored, inside and out, and is beautiful with its redwood staircase and ceiling, pendant trimmed lamps, Brussels carpet and rosewood piano. Many years ago this was a magnificent place, where names like Wyatt Erp, Curly Bill Brocius and Johnny Ringo were familiar. Shakespear was the town "too mean to live."

The rockhounding site shown on the map is reached by going about one mile south from Lordsburg to the fork at the cemetery. From there, bear right another one and four-tenths miles to the northern edge of the collecting area. At the given mileage, there are some ruts leading up the hill on the left which will take you to a few dumps of interest. Bright blue and green copper ores can be seen alongside this somewhat rough road, and some might be worth picking up.

There are eighty-five mines listed in the relatively condensed area surrounding the gravel road which stretches south from Lordsburg through the Pyramid Mountains. Most are abandoned, but a few aren't. Do not collect on private property. On the dumps of many abandoned mines, one can find fine specimens of bornite, malachite, azurite, chrysocolla, lead, linarite and countless other minerals of interest. Be patient and allow enough time to adequately explore the region.

Searching one of the dumps

Shakespear Area

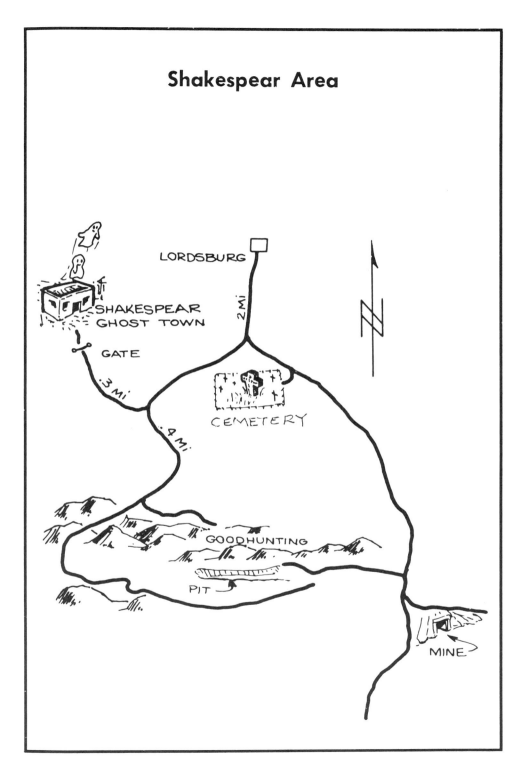

LORDSBURG

SHAKESPEAR
GHOST TOWN

GATE

2 Mi

.3 Mi

.4 Mi

CEMETERY

GOODHUNTING

PIT

MINE

ENGLE AREA

This location has been known by rockhounds for years, and because of that popularity, the beautiful red and orange carnelian that made it so famous is becoming tougher and tougher to find. There is still enough to make the trip worthwhile, though, if you are willing to spend some time patiently searching the hillsides.

Travel east from Truth or Consequences about seventeen miles on Highway 52 to the small town of Engle. The first five miles is very scenic as the road winds its way along the Rio Grande River past Elephant Butte Lake and through the Caballo Mountains. When you get to Engle, turn south on the ranch road and proceed eight miles to the remnants of Cutter and then another five miles to where you will see a large ranch house on the left about three-tenths of a mile from the road. This is the site of Aleman. At that point, follow the road leading to the right across the railroad tracks into the low hills as shown on the map.

The carnelian can be found throughout the terrain on both sides of this road for about two miles. It takes some patience and perseverance to find much, but frequently when only a bright red speck shows on the surface, it will be the "tip of the iceberg" with a much larger portion buried below.

There is some private ranch land in the vicinity, so be sure you don't trespass. Restrict all collecting to open land unless you gain permission beforehand.

There is another good carnelian area about one and one-half miles south of the Aleman site on the right across the railroad tracks in the low hills. If you have the time, be sure to explore it also.

Carnelian agate collected near Engle

Carnelian Agate

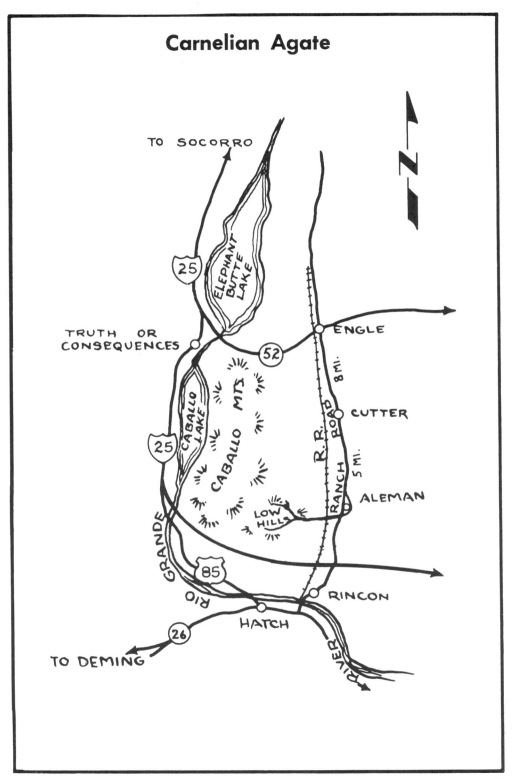

TO SOCORRO

25

ELEPHANT BUTTE LAKE

TRUTH OR CONSEQUENCES

52

ENGLE

CABALLO LAKE

CABALLO MTS

25

R.R. ROAD

RANCH

8 MI.

CUTTER

5 MI.

ALEMAN

LOW HILLS

GRANDE

85

RINCON

RIO

26

HATCH

TO DEMING

RIVER

RINCON AREA

Two locations, both near Rincon, offer collectors the opportunity to find lots of colorful jasper as well as fossilized coral. The jasper is scattered in random spots on the western slopes of the mountain with the large radio tower on top, while the fossils are in a small hill down below.

To get to the jasper site, follow the road leading east from town under Interstate 25 and up the hill. Be advised that once you start up the mountain it gets steep and there are few places to turn around; therefore, it is not recommended for large vehicles, or while towing a trailer.

Stop about half way to the top, park well off the road, and explore the western slopes. Be careful, however, since some of the soil is soft and the rocks unstable. The jasper chunks, scattered about in varying concentrations throughout the area, range in size from that of an egg to a large fist. Colors include brown, green and red, and most will take a good polish.

If you don't have much luck at the first stop, simply drive further along the road and try again. In addition to the colorful jasper, occasional pieces of agate, some filled with delicate dendrites, can also be found.

The fossil spot is reached by returning to the main road, at the base of the mountain, and locating the very faint ruts branching off to the east. If those ruts are severely washed out, it may be necessary to walk to the small fossil bearing hill located on the right not far from where the main road starts climbing toward the tower. The hill is made almost entirely of fossil coral, and some is suitable for cabinet display in or out of the host matrix.

Fossil specimens

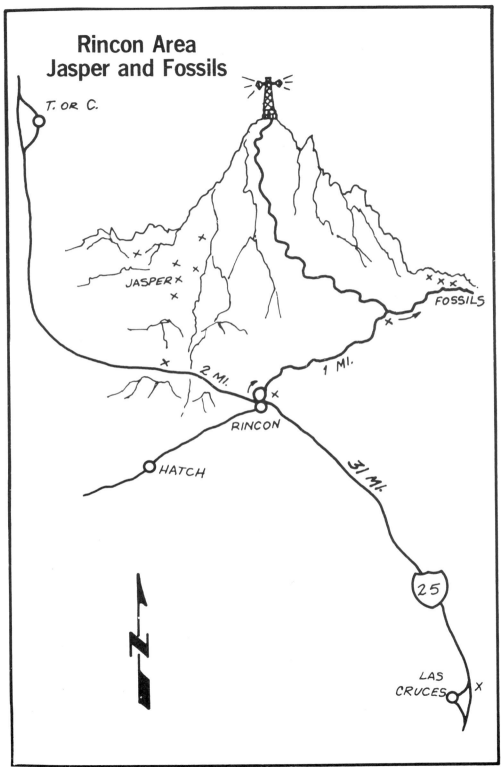

Rincon Area
Jasper and Fossils

T. OR C.

JASPER

FOSSILS

2 MI.

1 MI.

RINCON

HATCH

3I MI.

N

25

LAS CRUCES

RINCON QUARRY

Two collecting sites are very close to each other just off interstate 25 near the small town of Rincon. The first offers agate, jasper and quartz, while the other, an old abandoned quarry, provides an opportunity to find very well formed barite clusters.

Drive east from town under Interstate 25 for two-tenths of a mile to a crossroad. If you take the right fork, you will see some mounds continuing for quite a distance. It is in these little hills, Site A, where colorful cutting materials of agate and jasper are found. The agate, while generally small, is of good quality and occurs in such variety of colors and patterns. The jasper is primarily red, with some of the finest exhibiting yellow and orange flow patterns; however, it is not as plentiful as the agate.

After obtaining sufficient amounts of agate, jasper and quartz, head for Site B for excellent barite and calcite crystals from the dumps and walls of an old quarry. Take the left fork at the crossroad and go about four-tenths of a mile. This stretch is somewhat rough, so drive carefully or walk.

At the proper mileage, you will be ar the edge of the quarry. Park and begin searching the rubble scattered at the base of the steep walls. Look through all of the rocks and boulders to find excellent tabular crystals of barite, some measuring up to one-half inch across. By splitting seams, cavities can sometimes be found, revealing clusters of the prize crystals. More can be gathered by CAREFULLY removing chunks from the quarry wall, but be very cautious when climbing there or when working above your head with hard rock tools.

One might find fern-like pyrolusite dendrites on the host rock, nice brown banded rhyolite, and some low grade onyx, also.

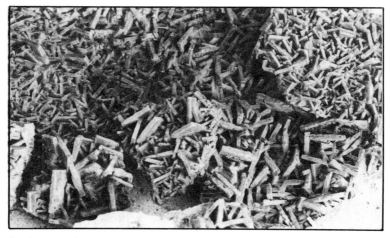

Barite crystals

Barite

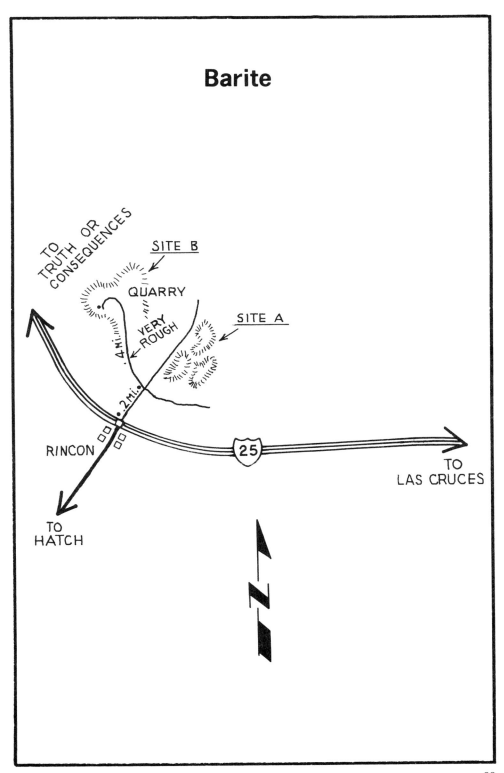

KILBOURNE HOLE

The Kilbourne Hole is one of the most interesting collecting areas in this book. It is an ancient crater, referred to as a "blowout," about one and one-fourth miles across and two hundred feet deep. The eruption which created this fascinating geological wonder occurred so many years ago that the shifting sand, so common in this part of New Mexico, has completely concealed its volcanic look and makes the crater appear more like an earthen dam.

The peridot is found in basalt nodular masses referred to as "bombs," sometimes measuring over a foot in diameter. When split in half, they are often filled with granular green masses of usually small peridot crystals. From time to time, however, gem quality specimens weighing several carats and measuring over one inch in diameter have been found here.

Most of what you will procure is small, but occasionally you might find a crystal large enough to facet. Carefully break up and separate the gem bearing "bombs," or meticulously search through the sand on the inside rim or floor of the crater for loose crystals.

Begin where previous rockhounds have dug into the contact zone inside the crater, just under the basalt, looking for the "bombs." The "bombs" are easy to identify, even though not overly plentiful on the surface. They contain few if any gas holes and look more like black petrified mud balls and blobs than the more common pumice of the crater.

Follow the instructions on the accompanying map to this site. The road isn't too bad but is sandy in places. Most rugged vehicles should have little problem, though. You will see the crater's large outer sand rim as you approach and must park at its base and hike over the crest to the inside on the loose sand. It is a difficult, but short, climb and the excitement of being able to collect in such a fascinating locality makes the effort worthwhile.

Parked at the base of Kilbourne crater

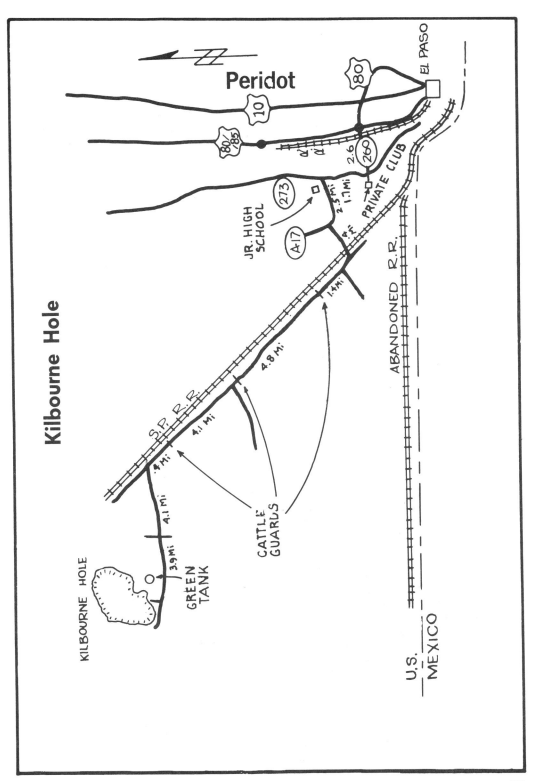

Kilbourne Hole

Peridot

HATCH AREA

This site offers rockhounds an opportunity to collect chalcedony roses, jasper and agate, as well as beautiful pink common opal, some of which is highly fluorescent. To get to this most productive location, head west one and one-tenth miles from Hatch on Highway 26 to County Road E-5. County Road E-5 intersects from the south just before you cross a bridge over a large wash and is easy to miss, so be looking for it as you approach the given mileage. Follow County Road E-5 south one and nine-tenths miles, turn left, and follow the ruts up and around the little hills. At that point, you will be in the center of the collecting area which extends throughout the hills and adjacent lowlands across the river bed to the east and onto the opposite banks. The best way to explore this site is to simply park and set out on foot.

The opal is found in seams on the west bank of the canyon, below the diversion dam, as shown on the map. The veins are somewhat small and occur in hard matrix, making it difficult to remove, but the vivid pink color produces beautiful cut stones.

The agate, jasper and chalcedony are found in float throughout the surrounding countryside in varying quantities and concentrations.

This is a fun place to explore with seemingly unlimited rockhounding opportunities. Local residents say that virtually all of the ranch roads in these hills will take you to additional jasper, agate and chalcedony rose collecting sites. It could be fun to spend a few days here to see if what they say is really true.

A sample of collectibles from Hatch area

Fluorescent Opal
and
Desert Roses

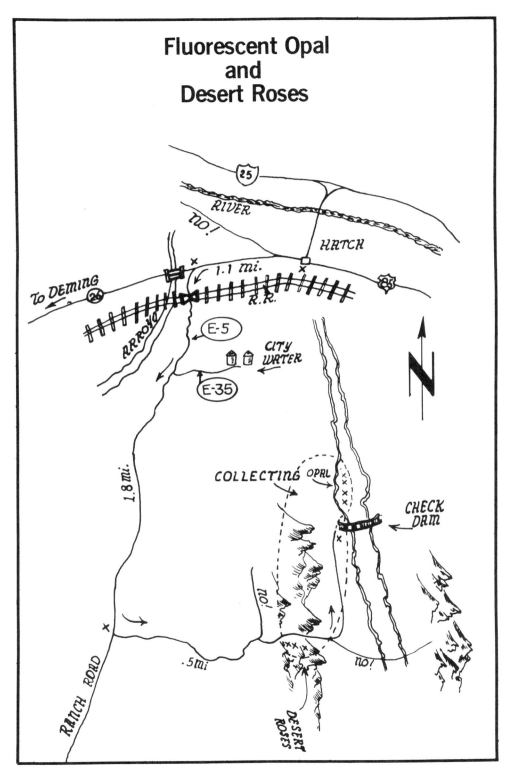

APACHE MINE #2

The dumps of the old Apache copper mine boast specimens of some of the most colorful minerals available to collectors, including bright green malachite, blue-green chrysocolla, and vivid blue azurite and turquoise.

To get to these highly productive dumps, take Highway 81 nineteen miles south from Interstate 10 to the small town of Hachita. Highway 81 joins Highway 9 in town, and at that point, you should go east about four-tenths of a mile and then turn right where Highway 81 branches off again to the south. Drive one and one-half miles to the cemetery and turn onto the dirt road seen branching off to the southeast.

Follow the main set of ruts, bearing right at the fork, as they lead around the western slopes of the small mountain range, as illustrated on the map. Four and one-half miles further on this rough dirt road will bring you to the old Apache Mine. At the time of publication, the mine was privately owned, but collectors were allowed to "trespass at their own risk." This status, of course, could be different when you visit. An inquiry in Hachita with Bill Bastian, the claimholder, might save you a long rough journey if there has been a change.

The dumps are filled with brilliantly colored copper ores which can be seen scattered all over the area even before you get out of your car. In addition to the beautiful minerals listed above, one can also obtain well formed rhombohedral calcite. Such specimens, when occurring with the colorful copper minerals, make beautiful cabinet pieces and are worth looking for.

Most of what is found here is specimen grade and too thin and/or crumbly to cut and polish, but lots of it is great for display in collections. Cutable material can be found here, though, so always be looking for it.

Parked near the Apache Mine dumps

Apache Mines

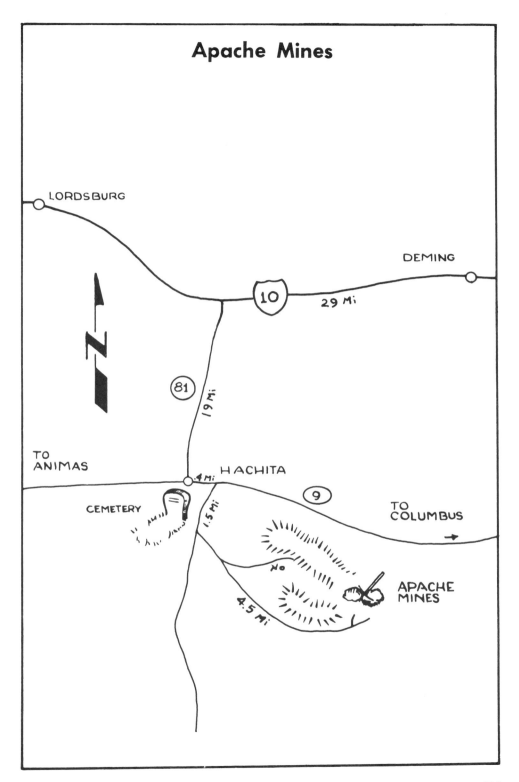

DUNKIN QUARRY

If you like to collect fossils and enjoy going to remote locations, be sure to visit the Dunkin Quarry situated thirty-seven and one-half miles west of Artesia. Embedded in the Permian limestone of the quarry, one can find outstanding specimens of fossilized mollusks, such as cephalopods and pelecypods, large clam shells and even a few trilobites.

To get there, drive west on Highway 82, four and four-tenths miles from where it intersects Highway 13. Then turn onto the ruts leading south. Continue about one more mile on this rugged dirt road to the quarry. This final stretch may not be suitable for all types of vehicles, so use good judgement before driving in. The hike is not too bad, and it would be better to walk than get stuck.

You will need gads, chisels and a sledge hammer to break up the hard limestone in an effort to expose the embedded fossils. This involves some hard work. The effort, however, is frequently rewarded with outstanding specimens in and out of matrix.

There is another more accessible fossil collecting site about sixty miles further west near the town of Cloudcroft. This is the Cloudcroft Quarry which is reached by going three-tenths of a mile north of town on Highway 24 as shown on the map. The quarry is on the left side of the road, a very short distance from the pavement, and the fossils are similar to those found at the Dunkin Quarry but not as plentiful.

If you continue west from Cloudcroft about seven miles, you will come to a tunnel and High Rolls Mountain Park. This is an excellent place to park and view Fresnal Canyon on your right. In addition, more fossils can be found in the surrounding rocks.

Fossil specimens from quarry

Fossils

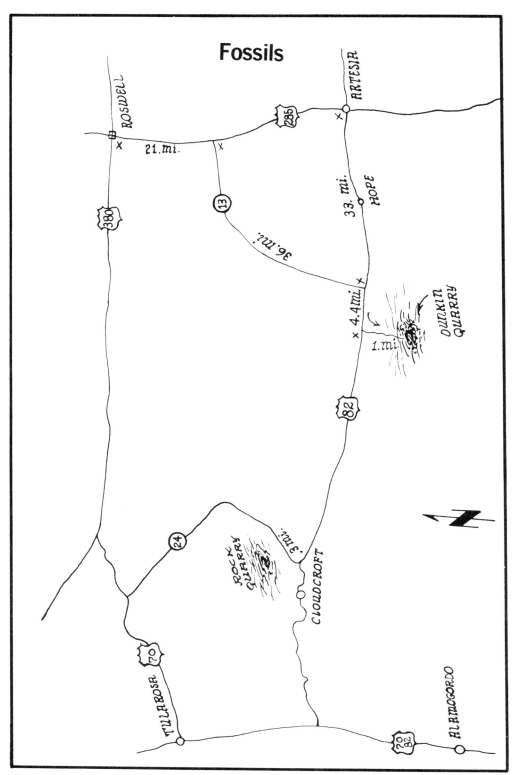

PECOS RIVER

Well formed, terminated quartz crystals known as Pecos Diamonds can be found all along the Pecos River, primarily between Fort Sumner in the north and Artesia in the south. The two localities mentioned here are especially productive and accessible. To get to the first, start in Roswell and head north on Highway 285 six miles. Turn east onto Highway 70 and continue to the Pecos River, another ten and eight-tenths miles. From there, proceed two and four-tenths more miles to the Bob Crosby Draw where some ruts will be seen heading from the pavement to the edge of the wash, as shown on the map. Either follow them, or simply park well off the highway and hike in.

The crystals, sometimes measuring over an inch in length, are found scattered throughout the wash for quite a distance in both directions. It is often helpful to use a screen to sift through the sand in the wash in order to help find otherwise buried crystals.

Another productive spot in the same general area is reached by continuing along Highway 70 another eight-tenths of a mile past Bob Crosby Draw to a ranch road which heads north. Take that road and, soon after leaving the highway, you will see lots of ruts going west toward some sand dunes. Turn onto any of them. You should be able to see the reddish crystal bearing soil through the thin vegetation. Look throughout these areas for the "diamonds." In addition to the Pecos Diamonds, one can also find pieces of petrified wood, agate and other collectibles.

The second prime Pecos Diamond location is reached by going four and one-half miles east of Artesia to the Pecos River. Continue a short distance further and take the first set of ruts leading south. Go about one-fourth of a mile, park, and search the area as you did at the Roswell sites. If you don't have much luck the first time, simply continue a little further and try again.

"Pecos Diamonds" loose, and in matrix

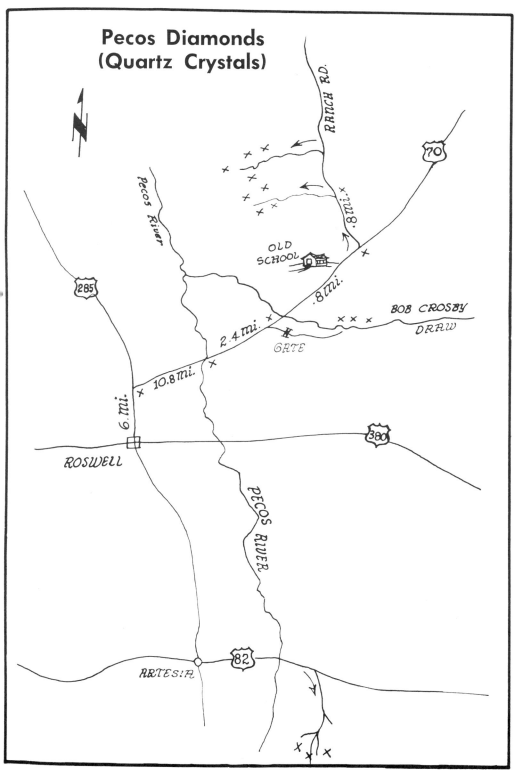

Pecos Diamonds
(Quartz Crystals)

RANCH RD.

70

Pecos River

285

OLD SCHOOL

.8 mi.

.8 mi.

BOB CROSBY DRAW

2.4 mi.

GATE

10.8 mi.

6. mi.

380

ROSWELL

PECOS RIVER

82

ARTESIA

Other Gem Trail books by publisher

Looking at the gem display at Rockhound State Park